Statistical Discovery.™ From SAS.

Version 11

Discovering JMP

"The real voyage of discovery consists not in seeking new
landscapes, but in having new eyes."

Marcel Proust

JMP, A Business Unit of SAS
SAS Campus Drive
Cary, NC 27513

11.1

The correct bibliographic citation for this manual is as follows: SAS Institute Inc. 2014. *Discovering JMP 11*®. Cary, NC: SAS Institute Inc.

Discovering JMP 11®

ISBN 978-1-61290-664-5

1st printing, September 2013

2nd printing, February 2014

Technology License Notices

- Scintilla - Copyright © 1998-2012 by Neil Hodgson <neilh@scintilla.org>.

 All Rights Reserved.

 Permission to use, copy, modify, and distribute this software and its documentation for any purpose and without fee is hereby granted, provided that the above copyright notice appear in all copies and that both that copyright notice and this permission notice appear in supporting documentation.

 NEIL HODGSON DISCLAIMS ALL WARRANTIES WITH REGARD TO THIS SOFTWARE, INCLUDING ALL IMPLIED WARRANTIES OF MERCHANTABILITY AND FITNESS, IN NO EVENT SHALL NEIL HODGSON BE LIABLE FOR ANY SPECIAL, INDIRECT OR CONSEQUENTIAL DAMAGES OR ANY DAMAGES WHATSOEVER RESULTING FROM LOSS OF USE, DATA OR PROFITS, WHETHER IN AN ACTION OF CONTRACT, NEGLIGENCE OR OTHER TORTIOUS ACTION, ARISING OUT OF OR IN CONNECTION WITH THE USE OR PERFORMANCE OF THIS SOFTWARE.

- Telerik RadControls: Copyright © 2002-2012, Telerik. Usage of the included Telerik RadControls outside of JMP is not permitted.

- ZLIB Compression Library - Copyright © 1995-2005, Jean-Loup Gailly and Mark Adler.

- Made with Natural Earth. Free vector and raster map data @ naturalearthdata.com.

- Packages - Copyright © 2009-2010, Stéphane Sudre (s.sudre.free.fr). All rights reserved.

 Redistribution and use in source and binary forms, with or without modification, are permitted provided that the following conditions are met:

 Redistributions of source code must retain the above copyright notice, this list of conditions and the following disclaimer.

 Redistributions in binary form must reproduce the above copyright notice, this list of conditions and the following disclaimer in the documentation and/or other materials provided with the distribution.

 Neither the name of the WhiteBox nor the names of its contributors may be used to endorse or promote products derived from this software without specific prior written permission.

 THIS SOFTWARE IS PROVIDED BY THE COPYRIGHT HOLDERS AND CONTRIBUTORS "AS IS" AND ANY EXPRESS OR IMPLIED WARRANTIES, INCLUDING, BUT NOT LIMITED TO, THE IMPLIED WARRANTIES OF MERCHANTABILITY AND FITNESS FOR A PARTICULAR PURPOSE ARE DISCLAIMED. IN NO EVENT SHALL THE COPYRIGHT OWNER OR CONTRIBUTORS BE LIABLE FOR ANY DIRECT, INDIRECT, INCIDENTAL, SPECIAL, EXEMPLARY, OR CONSEQUENTIAL DAMAGES (INCLUDING, BUT NOT LIMITED TO, PROCUREMENT OF SUBSTITUTE GOODS OR SERVICES; LOSS

Altered source versions must be plainly marked as such, and must not be misrepresented as being the original software.

The name of the author may not be used to endorse or promote products derived from this software without specific prior written permission.

- R software is Copyright © 1999-2012, R Foundation for Statistical Computing.

- MATLAB software is Copyright © 1984-2012, The MathWorks, Inc. Protected by U.S. and international patents. See www.mathworks.com/patents. MATLAB and Simulink are registered trademarks of The MathWorks, Inc. See www.mathworks.com/trademarks for a list of additional trademarks. Other product or brand names may be trademarks or registered trademarks of their respective holders.

Get the Most from JMP®

Whether you are a first-time or a long-time user, there is always something to learn about JMP.

Visit JMP.com to find the following:

- live and recorded webcasts about how to get started with JMP
- video demos and webcasts of new features and advanced techniques
- details on registering for JMP training
- schedules for seminars being held in your area
- success stories showing how others use JMP
- a blog with tips, tricks, and stories from JMP staff
- a forum to discuss JMP with other users

http://www.jmp.com/getstarted/

Contents
Discovering JMP

3 Visualizing Your Data
Graphing Data

4 Analyzing Your Data
Distributions, Relationships, and Models

Gallery of JMP Graphs
Various Graphs and Their Platforms

Here are pictures of many of the graphs that you can create with JMP. Each picture is labeled with the platform used to create it. For more information about the platforms and these and other graphs, see the documentation on the **Help > Books** menu.

Histogram
Analyze > Distribution

Bivariate
Analyze > Fit Y by X

Oneway
Analyze > Fit Y by X

Oneway t Test
Analyze > Fit Y by X

Logistic
Analyze > Fit Y by X

Mosaic Plot
Analyze > Fit Y by X

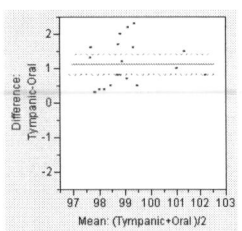

Matched Pairs
Analyze > Matched Pairs

Leverage Plot
Analyze > Fit Model

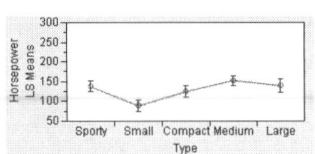

LS Means Plot
Analyze > Fit Model

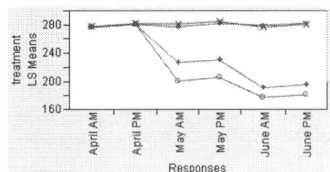

MANOVA
Analyze > Fit Model

Partition
Analyze > Modeling > Partition

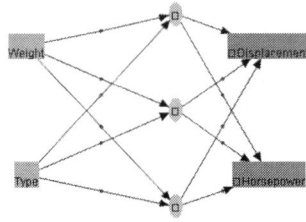

Neural Diagram
Analyze > Modeling > Neural

Actual by Predicted
Analyze > Modeling > Model Comparison

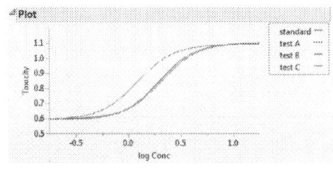

Nonlinear Fit
Analyze > Modeling > Nonlinear

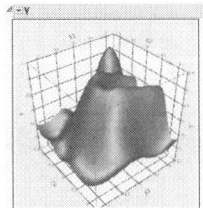

Surface Profiler
Analyze > Modeling > Gaussian Process

Time Series
Analyze > Modeling >
Time Series

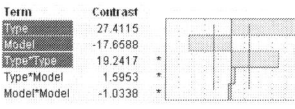

Screening
Analyze > Modeling >
Screening

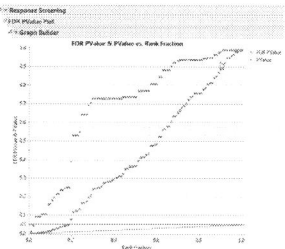

FDR pValue Plot
Analyze > Modeling >
Response Screening

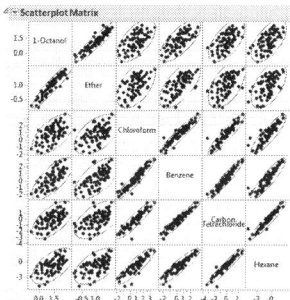

Scatterplot Matrix
Analyze >
Multivariate Methods >
Multivariate

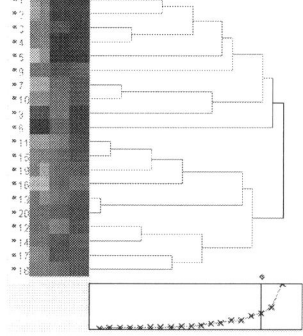

Dendrogram
Analyze >
Multivariate Methods >
Cluster

Self Organizing Map
Analyze >
Multivariate Methods >
Cluster

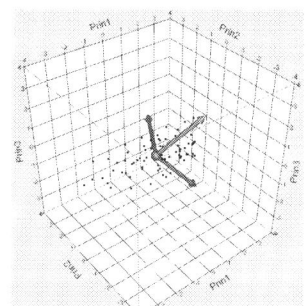

Principal Components
Analyze >
Multivariate Methods >
Principal Components

Canonical Plot
Analyze >
Multivariate Methods >
Discriminant

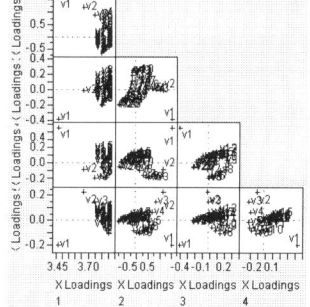

Loadings Plot
Analyze >
Multivariate Methods > Partial
Least Squares

XBar and R Charts
Analyze > Quality and Process
> Control Chart Builder

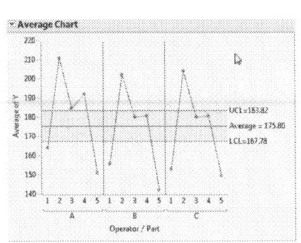

Average Chart
Analyze > Quality and Process
> Measurement Systems
Analysis

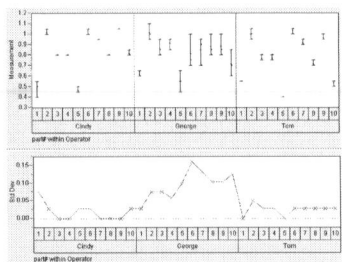

Variability Chart
Analyze > Quality and Process
> Variability/Attribute Chart

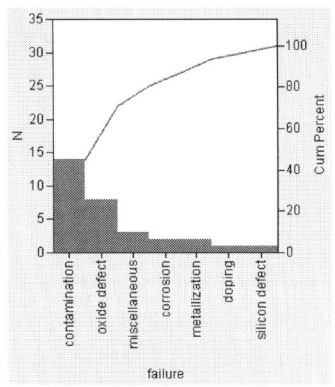

Goal Plot
Analyze > Quality and Process
> Capability

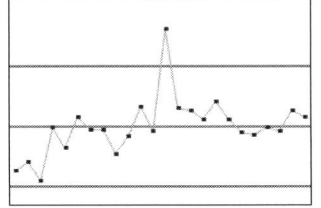

Individual Measurement Chart
Moving Range Chart
Analyze > Quality and Process
> Control Chart > IR

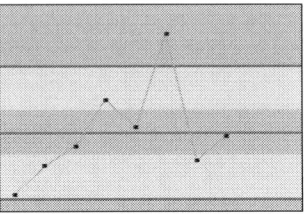

XBar Chart
Analyze > Quality and Process
> Control Chart > XBar

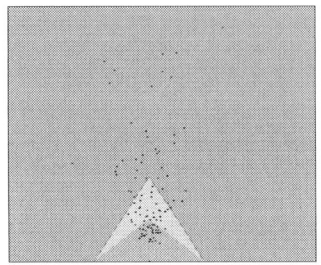

Pareto Plot
Analyze > Quality and Process
> Pareto Plot

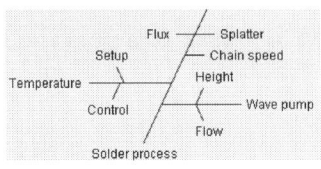

Ishikawa Chart
Fishbone Chart
Analyze > Quality and
Process> Diagram

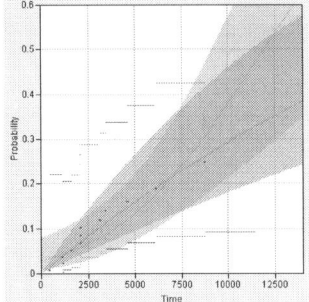

Compare Distributions
Analyze > Reliability and
Survival > Life Distribution

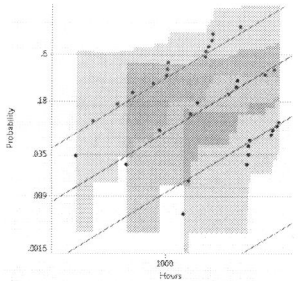

Nonparametric Overlay
Analyze > Reliability and
Survival > Fit Life by X

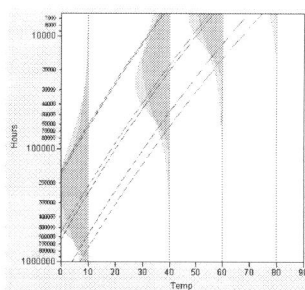

Scatterplot
Analyze > Reliability and
Survival > Fit Life by X

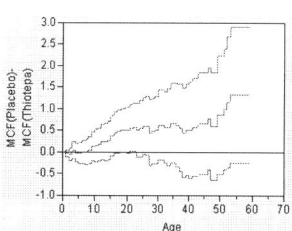

MCF Plot
Analyze > Reliability and
Survival > Recurrence
Analysis

Overlay
Analyze > Reliability and
Survival > Degradation

Forecast
Analyze > Reliability and
Survival > Reliability Forecast

Piecewise Weibull NHPP
Analyze > Reliability and
Survival > Reliability Growth

Reliability Block Diagram
Analyze > Reliability and
Survival > Reliability Block
Diagram

Failure Plot
Analyze > Reliability and
Survival > Survival

Survival Quantiles
Analyze > Reliability and
Survival > Fit Parametric
Survival

Baseline Survival
Analyze > Reliability and
Survival > Fit Proportional
Hazards

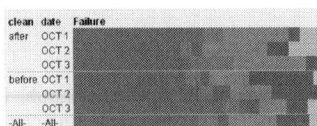

Mixture Profiler
Analyze > Consumer
Research > Categorical

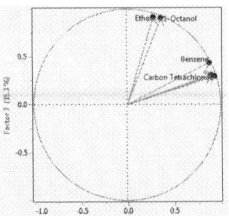

Factor Loading Plot
Analyze > Consumer
Research > Factor Analysis

Prediction Profile
Analyze > Consumer
Research > Choice

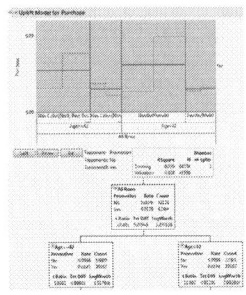

Uplift Model
Analyze > Consumer
Research > Uplift

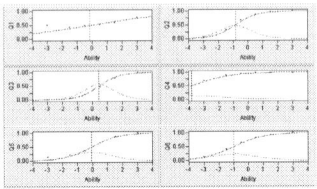

Characteristic Curves
Analyze > Consumer
Research > Item Analysis

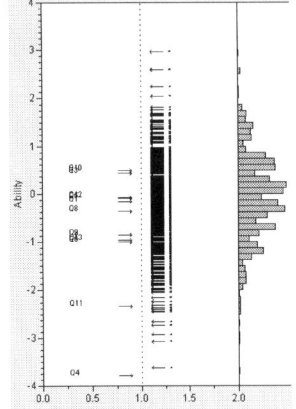

Dual Plot
Analyze > Consumer
Research > Item Analysis

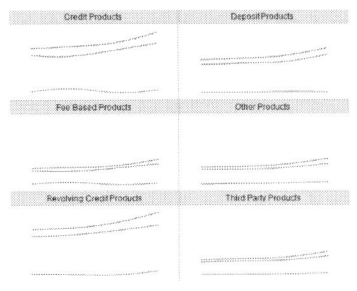

Line Graphs
Graph > Graph Builder

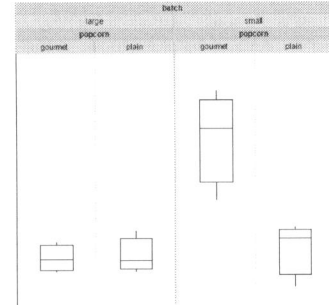

Box Plots
Graph > Graph Builder

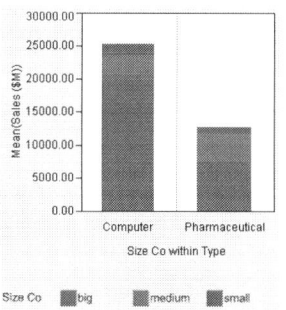

Stacked Bar Chart
Graph > Chart

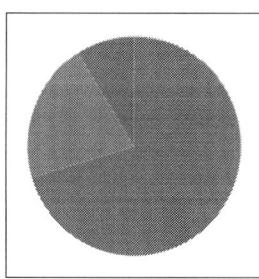

Pie Chart
Graph > Chart

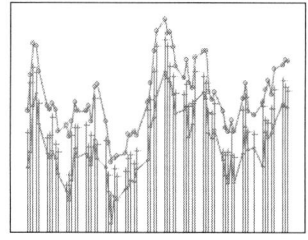

Needle and Line Chart
Graph > Overlay Plot

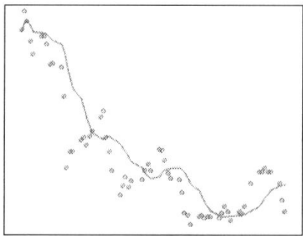

Dot and Line Chart
Graph > Overlay Plot

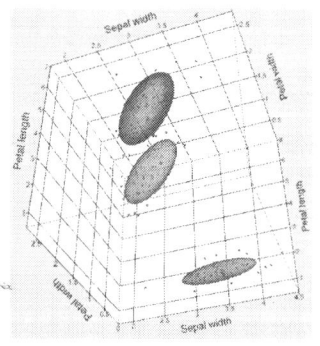

Three Dimensional Scatterplot
Graph > Scatterplot 3D

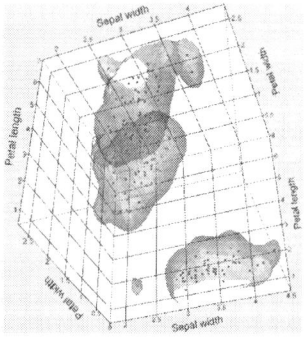

Three Dimensional Scatterplot
Graph > Scatterplot 3D

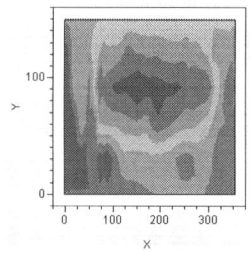

Contour Plot
Graph > Contour Plot

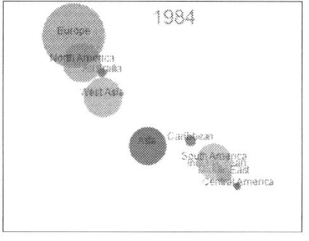

Bubble Plot
Graph > Bubble Plot

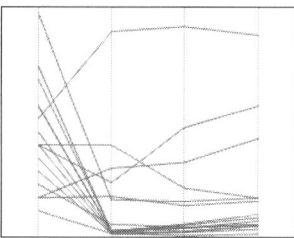

Parallel Plot
Graph > Parallel Plot

Cell Plot
Graph > Cell Plot

Treemap
Graph > Treemap

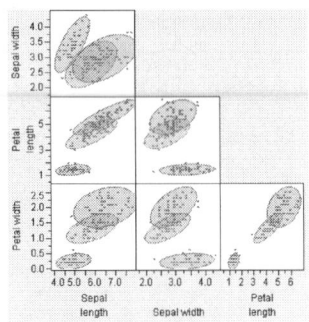

Scatterplot Matrix
Graph > Scatterplot Matrix

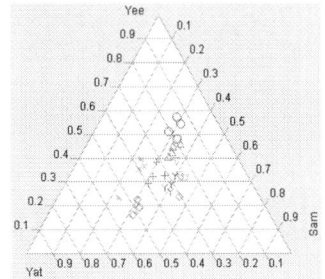

Ternary Plot
Graph > Ternary Plot

Prediction Profiler
Graph > Profiler

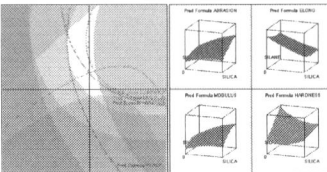

Contour Profiler
Graph > Contour Profiler

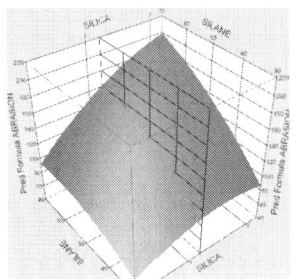

Surface Plot
Graph > Surface Plot

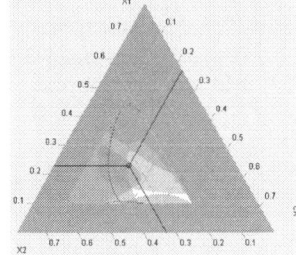

Mixture Profiler
Graph > Mixture Profiler

About This Guide

Discovering JMP provides a general introduction to the JMP software. This guide assumes that you have no knowledge of JMP. Whether you are an analyst, researcher, student, professor, or statistician, this guide gives you a general overview of JMP's user interface and features.

This guide introduces you to the following information:

- Starting JMP
- The structure of a JMP window
- Preparing and manipulating data
- Using interactive graphs to learn from your data
- Performing simple analyses to augment graphs
- Customizing JMP and special features

This guide contains six chapters. Each chapter contains examples that reinforce the concepts presented in the chapter. All of the statistical concepts are at an introductory level. The sample data used in this book are included with the software. Here is a description of each chapter:

- Chapter 1, "Introducing JMP", provides an overview of the JMP application. You learn how content is organized and how to navigate the software.
- Chapter 2, "Working with Your Data", describes how to import data from a variety of sources, and prepare it for analysis. There is also an overview of data manipulation tools.
- Chapter 3, "Visualizing Your Data", describes graphs and charts that you can use to visualize and understand your data. The examples range from simple analyses involving a single variable, to multiple-variable graphs that enable you to see relationships between many variables.
- Chapter 4, "Analyzing Your Data", describes many commonly used analysis techniques. These techniques range from simple techniques that do not require the use of statistical methods, to advanced techniques, where knowledge of statistics is useful.
- Chapter 5, "Saving and Sharing Your Work", describes using journals and projects, and saving scripts.
- Chapter 6, "Special Features", describes how to automatically update graphs and analyses as data changes, how to use preferences to customize your reports, and how JMP interacts with SAS.

After reviewing this guide, you will be comfortable navigating and working with your data in JMP.

JMP is available for both Windows and Macintosh operating systems. However, the material in this guide is based on a Windows operating system.

Chapter 1

Introducing JMP
Basic Concepts

JMP (pronounced jump) is a powerful and interactive data visualization and statistical analysis tool. Use JMP to learn more about your data by performing analyses and interacting with the data using data tables, graphs, charts, and reports.

JMP is useful to the researcher who wants to perform a wide range of statistical analyses and modeling. JMP is equally useful to the business analyst who wants to quickly uncover trends and patterns in data. With JMP, you do not have to be an expert in statistics to get information from your data.

For example, you can use JMP to do the following:

- Create interactive graphs and charts to explore your data and discover relationships.
- Discover patterns of variation across many variables at once.
- Explore and summarize large amounts of data.
- Develop powerful statistical models to predict the future.

Figure 1.1 Examples of JMP Reports

Contents

Concepts That You Should Know

Before you begin using JMP, you should be familiar with these concepts:

- Enter, view, edit, and manipulate data using JMP *data tables*.

- Select a *platform* from the **Analyze** and **Graph** menus. Platforms contain interactive windows that you use to analyze data and work with graphs.

- Platforms use these windows:

 - *Launch windows* where you set up and run your analysis.

 - *Report windows* showing the output of your analysis.

- Report windows normally contain the following items:

 - A graph of some type (such as a scatterplot or a chart).

 - Specific *reports* that you can show or hide using the *disclosure button* ◢ .

 - Platform *options* that are located within *red triangle menus* ▼ .

How Do I Get Started?

The general workflow in JMP is simple:

1. Get your data into JMP.

2. Select a platform and complete its launch window.

3. Explore your results and discover where your data takes you.

This workflow is described in more detail in "Understanding the JMP Workflow" on page 28.

Typically, you start your work in JMP by using graphs to visualize individual variables and relationships among your variables. Graphs make it easy to see this information, and to see the deeper questions to ask. Then you use analysis platforms to dig deeper into your problems and find solutions.

- The "Working with Your Data" chapter on page 37 shows you how to get data into JMP.

- The "Visualizing Your Data" chapter on page 67 shows you how to use some of the useful graphs JMP provides to look more closely at your data.

- The "Analyzing Your Data" chapter on page 99 shows you how to use some of the analysis platforms.

Each chapter teaches through examples. The following sections in this chapter describe data tables and general concepts for working in JMP.

Starting JMP

Start JMP in two ways:

- Double-click on the JMP icon, normally found on your desktop. This starts JMP, but does not open any existing JMP files.

- Double-click an existing JMP file. This starts JMP and opens the file.

The initial view of JMP includes the Tip of the Day window and the Home window on Windows; on Macintosh, the Tip of the Day and JMP Starter window initially appear.

The JMP Starter window classifies actions and platforms using categories.

Figure 1.2 The JMP Starter

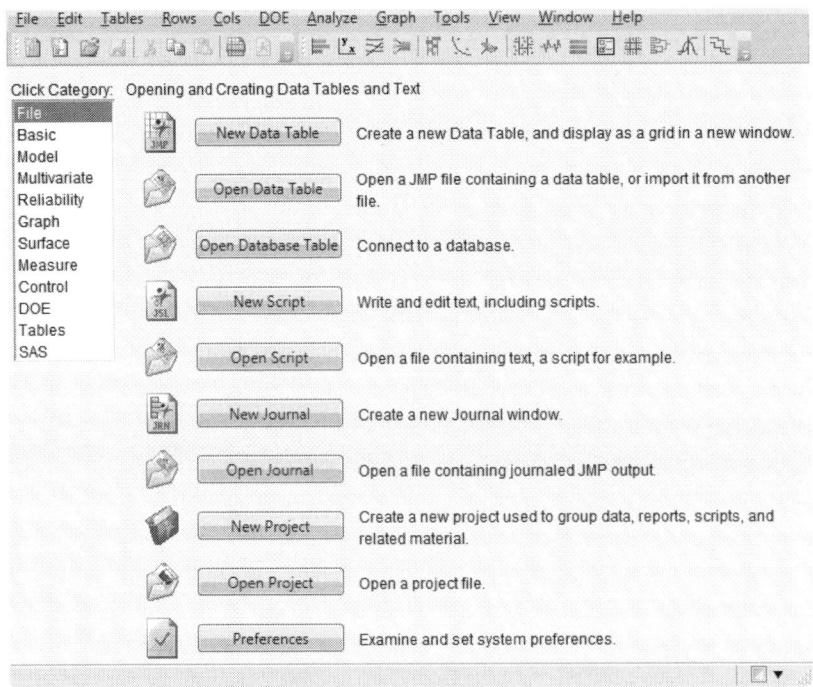

On the left is a list of categories. Click a category to see the features and the commands related to that category. For a description of all of the features in the JMP Starter, see *Using JMP*.

On Windows, another useful window is the Home window.

Figure 1.3 The Home Window (Windows Only)

To open the Home window, select **View > Home Window**. This window includes links to the following:

- the data tables and report windows that are currently open
- files that you have opened recently

For more details about the JMP Starter window and the Home window, see *Using JMP*.

Almost all JMP windows contain a menu bar and a toolbar. You can find most JMP features in three ways:

- using the menu bar
- using the toolbar buttons
- using the buttons on the JMP Starter window

Note: By default, windows in JMP are not maximized. This enables you to see the interaction between the windows.

About the Menu Bar and Toolbars

The menus and toolbars are hidden in many windows. To see them, hover your mouse cursor over the blue bar under the window's title bar. The menus in the JMP Starter window, the Home window, and all data tables are always visible.

Using Sample Data

The examples in this book and the other JMP books use sample data tables. The default location on Windows for the sample data is:

 C:\Program Files\SAS\JMP\<version>\Samples\Data

The Sample Data Index groups the data tables by category. Click a disclosure button to see a list of data tables for that category, and then click a link to open a data table.

Macintosh sample data is usually installed at the root level at:

 /Library/Application Support/JMP/<version>/Samples/Data

Opening a JMP Sample Data Table

1. From the **Help** menu, select **Sample Data**.
2. Open the **Data Tables used in Discovering JMP** list by clicking on the disclosure button next to it.
3. Click the name of the data table to use it in the examples in this book.

Sample Import Data

Use files from other applications to learn how to import data into JMP.

The default location on Windows for the sample import data is:

 C:\Program Files\SAS\JMP\<version>\Samples\Import Data

Understanding Data Tables

A data table is a collection of data organized in rows and columns. It is similar to a Microsoft® Excel® spreadsheet, but with some important differences that are discussed in "How is JMP Different from Excel?" on page 33. A data table might also contain other information like notes, variables, and scripts. These supplementary items are discussed in later chapters.

Open the VA Lung Cancer data table to see the data table described here.

Figure 1.4 A Data Table

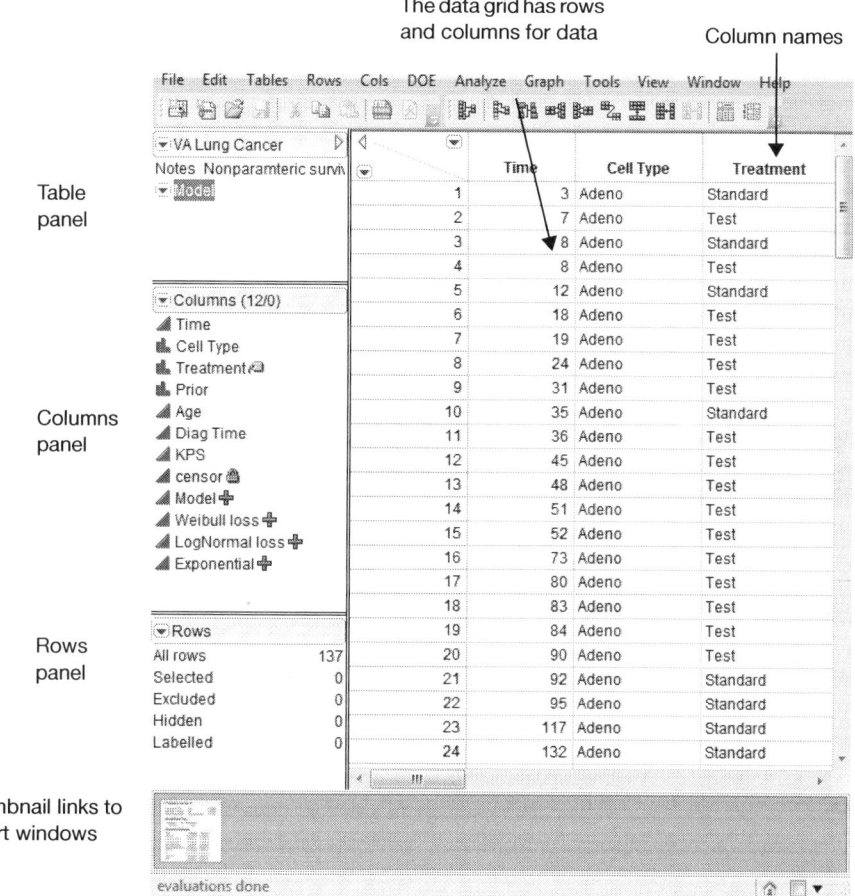

A data table contains the following parts:

Data grid The data grid contains the data arranged in rows and columns. Generally, each row
in the data grid is an observation, and the columns (also called variables) give information
about the observations. In Figure 1.4, each row corresponds to a test subject, and there are
twelve columns of information. Although all twelve columns cannot be shown in the data
grid, the Columns panel lists them all. The information given about each test subject
includes the time, cell type, treatment, and more. Each column has a header, or name. That
name is not part of the table's total count of rows.

Table panel The table panel can contain table variables or table scripts. In Figure 1.4, there is
one saved script called **Model** that can automatically recreate an analysis. This table also
has a variable named Notes that contains information about the data. Table variables and
table scripts are discussed in a later chapter.

Columns panel The columns panel shows the total number of columns, whether any columns are selected, and a list of all the columns by name. The numbers in parentheses (12/0) show that there are twelve columns, and that no columns are selected. An icon to the left of each column name shows that column's modeling type. Modeling types are described in "Understanding Modeling Types" on page 103 in the "Analyzing Your Data" chapter. Icons to the right show any attributes assigned to the column. See "Viewing or Changing Column Information" on page 51 in the "Working with Your Data" chapter for more information about these icons.

Rows panel The rows panel shows the number of rows in the data table, and how many rows are selected, excluded, hidden, or labeled. In Figure 1.4, there are 137 rows in the data table.

Thumbnail links to report windows This area shows thumbnails of all reports based on the data table. Hover over a thumbnail to see a larger preview of the report window. Double-click a thumbnail to bring the report window to the front.

Interacting with the data grid, which includes adding rows and columns, entering data, and editing data, is discussed in the "Working with Your Data" chapter on page 37. If you open multiple data tables, each one appears in a separate window.

Understanding the JMP Workflow

Once your data is in a data table, you can create graphs or plots, and perform analyses. All features are located in platforms, which are found primarily on the **Analyze** or **Graph** menus. They are called platforms because they do not just produce simple static results. Platform results appear in report windows, are highly interactive, and are linked to the data table and to each other.

The platforms under the **Analyze** and **Graph** menus provide a variety of analytical features and data exploration tools.

The general steps to produce a graph or analysis are as follows:

1. Open a data table.
2. Select a platform from the Graph or Analysis menu.
3. Complete the platform launch window to set up your analysis.
4. Click OK to create the report window that contains your graphs and statistical analyses.
5. Customize your report by using report options.
6. Save, export, and share your results with others.

Later chapters discuss these concepts in greater detail.

The following example shows you how to perform a simple analysis and customize it in four steps. This example uses the Companies.jmp file sample data table to show a basic analysis of the variable Profits ($M).

Step 1: Launching a Platform and Viewing Results

1. Open the Companies.jmp data table.

2. Select **Analyze > Distribution** to open the Distribution launch window.

3. Select Profits ($M) in the Select Columns box and click the **Y, Columns** button.

 The variable Profits ($M) appears in the **Y, Columns** role. See Figure 1.5 for the completed window.

 Another way to assign variables is to click and drag columns from the Select Columns box to any of the roles boxes.

Figure 1.5 Assign Profits ($M)

4. Click **OK**.

 The Distribution report window appears.

Figure 1.6 Distribution Report Window

The report window contains basic plots or graphs and preliminary analysis reports. The results appear in an outline format, and you can show or hide any report by clicking on the disclosure button.

Red triangle menus contain options and commands to request additional graphs and analyses at any time.

- Hover over the blue bar at the top of the window to see the menu bar and the toolbars.
- Click the data table button to bring the data table that was used to create this report to the front.
- Click the Home window button to return to the Home window.

Step 2: Removing the Box Plot

Continue using the Distribution report that you created earlier.

1. Click the red triangle next to Profits ($M) to see a menu of report options.
2. Deselect **Outlier Box Plot** to turn the option off.

 The outlier box plot is removed from the report window.

Figure 1.7 Removing the Outlier Box Plot

Click here to remove the outlier box plot.

Step 3: Requesting Additional Output

Continue to use the same report window.

1. From the red triangle menu next to Profits ($M), select **Test Mean**.

 The Test Mean window appears.

2. Enter 500 in the **Specify Hypothesized Mean** box.

3. Click **OK**.

 The test for the mean is added to the report window.

Figure 1.8 Test for the Mean

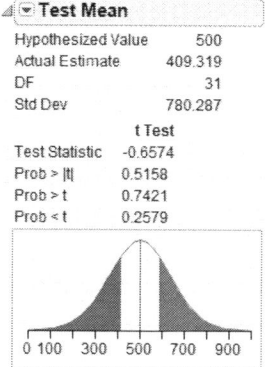

Step 4: Interacting with Platform Results

All platforms produce results that are interactive. For example:

- Reports can be shown or hidden.
- Additional graphs and statistical details can be added or removed to suit your purposes.
- Platform results are connected to the data table and to each other.

For example, to close the **Quantiles** report, click the disclosure button next to **Quantiles**.

Figure 1.9 Close the Quantiles Report

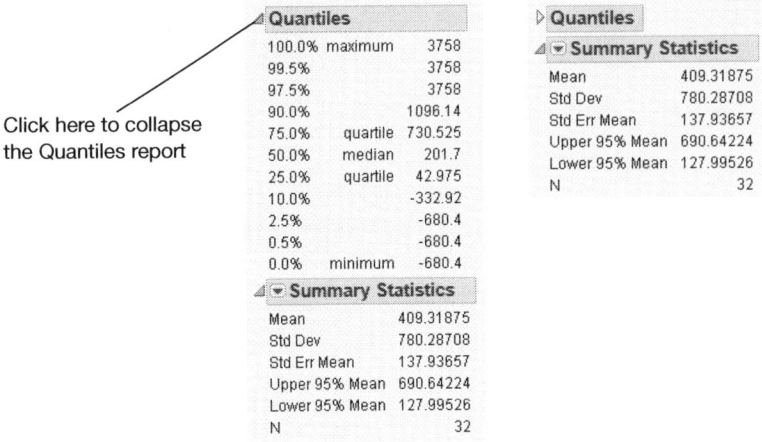

Click here to collapse
the Quantiles report

Platform results are connected to the data table. The histogram in Figure 1.10 shows that a group of companies makes a much higher profit that the others. To quickly identify that group, click on the histogram bar for them. The corresponding rows in the data table are selected.

Figure 1.10 Connection Between Platform Results and Data Table

Click the bar to select the corresponding rows

In this case, the group includes only one company, and that one row is selected.

How is JMP Different from Excel?

There are a number of important differences between JMP and Excel or other spreadsheet applications.

Table 1.1 JMP and Excel Differences

Formulas	
Excel	Formulas are applied to individual cells.
JMP	Formulas are applied only to entire columns. "Calculating Values with Formulas" on page 52 in the "Working with Your Data" chapter describes how to use formulas.
Column Names	
Excel	Column names are part of the grid. Numbered rows and labeled columns extend past the data. Numeric and character data reside in the same column.
JMP	Column names are not part of the grid. There are no rows and columns beyond the existing data. The grid is only as big as the data. A column is either numeric or character. If a column contains both character and numeric data, the entire column's data type is character, and the numbers are treated as character data.
	"Understanding Modeling Types" on page 103 in the "Analyzing Your Data" chapter describes how data type influences platform results.
Tables and Worksheets	
Excel	A single spreadsheet contains several tables, or worksheets.
JMP	JMP does not have the concept of worksheets. Each data table is a separate .jmp file and appears in a separate window.
Date Grid	
Excel	Data can be located anywhere in the data grid.
JMP	Data always begins in row 1 and column 1.
Analysis and Graph Reports	

Table 1.1 JMP and Excel Differences

Excel	All data, analyses, and graphs are placed inside the data grid.
JMP	Results appear in a separate window.

How Do I Find Help?

As you start using JMP, a variety of resources are available to supplement your learning. The JMP books, Help, Web information, sample data, and more are all available on the Help menu.

Table 1.2 Help Options

Option	Description
Help Contents, Search the Help, and Help Index	These three menu items open the JMP Help system. The Help system provides navigable and searchable documentation.
Books	Contains links to the full documentation book set, a description of all the JMP menus, and a quick reference describing keyboard shortcuts.
New Features	Document detailing new features for the current version.
JMP User Community	Takes you to the online user forum on the JMP Web site (www.jmp.com) where you can download JMP files, read the JMP Blog, and discuss JMP topics with other users. The Web site contains information about JMP and provides links to latest news and events, and file sharing. You can register for free weekly webinars, and find podcasts and other JMP literature. The Web site also contains videos that demonstrate JMP features. You can also learn how to join one of the regional JMP User's Groups.
Tutorials	Interactive tutorials that demonstrate how to use some of JMP's statistical and graphical features.
	If you are not familiar with JMP, start with the Beginners Tutorial. Every new JMP user should take this five-minute tutorial on the user-interface basics of JMP.
Sample Data	Provides links to all the sample data used in the documentation. The sample data tables help when you are learning JMP.
Tip of the Day	A collection of helpful tips and hints that enhance your experience with JMP.
Statistics Index	Provides definitions of statistical terms.

Table 1.2 Help Options *(Continued)*

Option	Description
Scripting Index	Describes JSL objects and messages and provides examples and help.
About JMP	Shows the release, copyright, processing (32- or 64-bit), operating system, and the owner of the copy of JMP that is running.

Chapter **2**

Working with Your Data
Preparing Your Data for Graphing and Analyzing

Before graphing or analyzing your data, the data has to be in a data table and in the proper format. This chapter shows some basic data management tasks, including the following:

- Creating new data tables
- Opening existing data tables
- Importing data from other applications into JMP
- Managing your data

Figure 2.1 Example of a Data Table

	Type	Size Co	Sales ($M)	Profits ($M)	# Employ	profit/emp	Assets	%pf
1	Computer	small	855.1	31.0	7523	4120.70	615.2	
2	Pharmaceutical	big	5453.5	859.8	40929	21007.11	4851.6	
3	Computer	small	2153.7	153.0	8200	18658.54	2233.7	
4	Pharmaceutical	big	6747.0	1102.2	50816	21690.02	5681.5	
5	Computer	small	5284.0	454.0	12068	37620.15	2743.9	
6	Pharmaceutical	big	9422.0	747.0	54100	13807.76	8497.0	
7	Computer	small	2876.1	333.3	9500	35084.21	2090.4	
8	Computer	small	709.3	41.4	5000	8280.00	468.1	
9	Computer	small	2952.1	-680.4	18000	-37800.0	1860.7	
10	Computer	small	784.7	89.0	4708	18903.99	955.8	
11	Computer	small	1324.3	-119.7	13740	-8711.79	1040.2	
12	Pharmaceutical	medium	4175.6	939.5	28200	33315.60	5848.0	
13	Computer	big	11899.0	829.0	95000	8726.32	10075.0	
14	Computer	small	873.6	79.5	8200	9695.12	808.0	
15	Pharmaceutical	big	9844.0	1082.0	83100	13020.46	7919.0	
16	Pharmaceutical	small	969.2	227.4	3418	66530.13	784.0	
17	Pharmaceutical	medium	6698.4	1495.4	34400	43470.93	6756.7	
18	Computer	big	5956.0	412.0	56000	7357.14	4500.0	
19	Pharmaceutical	big	5903.7	681.1	42100	16178.15	8324.8	
20	Computer	medium	2959.3	252.8	31404	8049.93	5611.1	
21	Pharmaceutical	small	1198.3	86.5	8527	10144.25	1791.7	
22	Computer	small	900.5	20.0	8579	2426.47	624.3	

Companies panel:
- Locked File C:\Program Files\S/
- Notes Selected Data on the Fort

Columns (8/0)
- Type
- Size Co
- Sales ($M)
- Profits ($M)
- # Employ
- profit/emp
- Assets
- %profit/sales

Rows
All rows	32
Selected	0
Excluded	0
Hidden	0
Labelled	0

Contents

Getting Your Data into JMP

- To copy and paste data from another application, see "Copying and Pasting Data" on page 39.

- To import data from another application, see "Importing Data" on page 39.

- To enter data directly into a data table, see "Entering Data" on page 43

- To open a data table, double-click on the file, or use the **File > Open** command.

You can also import data into JMP from a database. For more information, see *Using JMP*.

This chapter uses sample data tables and sample import data that is installed with JMP. To find these files, see "Using Sample Data" on page 26 in the "Introducing JMP" chapter.

Copying and Pasting Data

You can move data into JMP by copying and pasting from another application, such as Excel or a text file.

Example: Copying and Pasting Data

1. Open the VA Lung Cancer.xls file in Excel. This file is located in the Sample Import Data folder.

2. Select all of the rows and columns, including the column names. There are 12 columns and 138 rows.

3. Copy the selected data.

4. In JMP, select **File > New > Data Table** to create an empty table.

5. Select **Edit > Paste with Column Names** to paste the data and column headings.

 If the data that you are pasting into JMP does *not* have column names, then you can use **Edit > Paste**.

Importing Data

You can move data into JMP by importing data from another application, such as Excel, SAS, or text files. The basic steps to import data are as follows:

1. Select **File > Open**.

2. Navigate to your file's location.

3. If your file is not listed in the Open Data File window, select the correct file type from the **Files of type** menu.

4. Click **Open**.

Example: Importing a Microsoft Excel File

1. Select **File > Open**.

2. Navigate to the Samples/Import Data folder.

3. Select Team Results.xls.

 Note the rows and columns on which the data begin. The spreadsheet also contains two worksheets. In this example, you import the Ungrouped Team Results worksheet.

4. Click **Open**.

 The spreadsheet opens in the Excel Import Wizard, where a preview of the data appears along with import options.

 Text from the first row of the spreadsheet are column headings. However, you want text in row 3 of the spreadsheet to be converted to column headings.

5. Next to **Column headers start on row**, type 3 and press **Enter**. The column headings are updated in the data preview. The value for the first row of data is updated to 4.

6. Save the settings only for this worksheet:

 – Deselect **Use for all worksheets** in the lower left corner of the window.

 – Select **Ungrouped Team Results** in the upper right corner of the window.

7. Click **Import** to convert the spreadsheet as you specified.

When you import Excel files, JMP predicts whether columns headings exist, and if the column names are on row one. The copy and paste method is recommended for the following situations:

* If the column names are located in a row other than row one

* If the file does not include column names and the data does not start in row one

* If the file contains column names and the data does not start in row two

See "Copying and Pasting Data" on page 39 and *Using JMP* for more information about importing Excel files.

Example: Importing a Text File

One way to import a text file is to let JMP assume the data's format and place the data in a data table. This method uses settings that you can specify in Preferences. See *Using JMP* for information about setting text import preferences.

Another way to import a text file is to use a Text Preview window to see what your data table will look like after importing, and make adjustments. The following example shows you how to use Text Import Preview window.

1. Select **File > Open**.

2. Navigate to the Samples/Import Data folder.

3. Select Animals_line3.txt.

4. At the bottom of the Open window, select **Data with Preview**.

5. Click **Open**.

Figure 2.2 Initial Preview Window

This text file has a title on the first line, column names on the third line, and the data starts on line four. If you opened this directly in JMP, the Animals Data line would be the first column name, and all the column names and data afterward would be out of sync. The Preview window lets you adjust the settings before you open the file, and see how your adjustments affect the final data table.

6. Enter 3 in the **File contains column names on line** field.

7. Enter 4 in the **Data starts on line** field.

8. Click **Next**.

 In the second window, you can exclude columns from the import and change the data modeling of the columns. For this example, use the default settings.

9. Click **Import**.

The new data table has columns named species, subject, miles, and season. The species and season columns are character data. The subject and miles columns are continuous numeric data.

Transfer Data from Excel

You can use the JMP Add In for Excel to transfer a spreadsheet from Excel to JMP:

- a data table
- Graph Builder
- Distribution platform
- Fit Y by X platform
- Fit Model platform
- Time Series platform
- Control Chart platform

Set JMP Add In Preferences in Excel

To configure JMP Add In Preferences:

1. In Excel, select **JMP > Preferences**.

 The JMP Preferences window appears.

Figure 2.3 JMP Add In Preferences

2. Accept the default **Data Table Name** (File name_Worksheet name) or type a name.
3. Select to **Use the first rows as column names** if the first row in the worksheet contains column headers.
4. If you selected to use the first rows a column headers, type the number of rows used.
5. Select to **Transfer Hidden Rows** if the worksheet contains hidden rows to be included in the JMP data table..

6. Select to **Transfer Hidden Column** if the worksheet contains hidden columns to be included in the JMP data table.

7. Click **OK** to save your preferences.

Transfer to JMP

To transfer an Excel worksheet to JMP:

1. Open the Excel file.

2. Select the worksheet to transfer.

3. Select **JMP**, then select the JMP destination:

 - Data Table
 - Graph Builder
 - Distribution platform
 - Fit Y by X platform
 - Fit Model platform
 - Time Series platform
 - Control Chart platform

 The Excel worksheet is opened as a data table in JMP and the selected platform's launch window appears.

Entering Data

You can enter data directly in a data table. The following example shows you how to enter data that was collected over several months into a data table.

Scenario

Table 2.1 shows the data from a study that investigated a new blood pressure medication. Each individual's blood pressure was measured over a six-month period. Two doses (300mg and 450mg) of the medication were used, along with a control and placebo group. The data shows the average blood pressure for each group.

Table 2.1 Blood Pressure Data

Month	Control	Placebo	300mg	450mg
March	165	163	166	168
April	162	159	165	163
May	164	158	161	153

Table 2.1 Blood Pressure Data *(Continued)*

Month	Control	Placebo	300mg	450mg
June	162	161	158	151
July	166	158	160	148
August	163	158	157	150

Entering Data in a New Data Table

1. Select **File > New > Data Table** to create an empty data table.

 A new data table has one column and no rows.

2. Click the column name to select it, and then click again to edit the name.

Note: If you double-click too quickly, the Column Info window appears. You can change the column name there as well.

3. Change the column name to Month. See Figure 2.4.

Figure 2.4 Entering a Column Name

 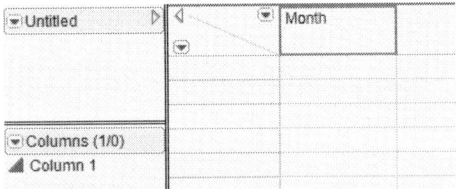

Click once to select the column. Click again, and then type "Month".

4. Select **Rows > Add Rows**.

 The Add Rows window appears.

5. Since you want to add six rows, type 6.

6. Click **OK**. Six empty rows are added to the data table.

7. Enter the Month information by double-clicking in a cell and typing.

Figure 2.5 Month Column Completed

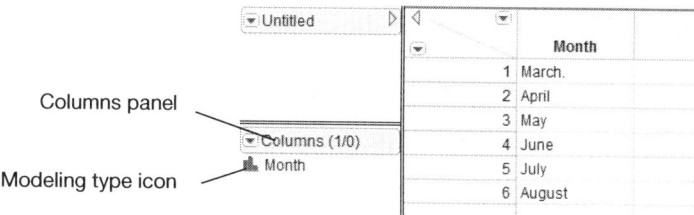

Columns panel

Modeling type icon

In the columns panel, look at the modeling type icon to the left of the column name. It has changed to reflect that Month is now nominal (previously it was continuous). Compare the modeling type shown for Column 1 in Figure 2.4 and for Month in Figure 2.5. This difference is important and is discussed in "Viewing or Changing Column Information" on page 51.

8. Double-click in the space on the right side of the Month column to add the Control column.

9. Change the name to Control.

10. Enter the Control data as shown in Table 2.1. Your data table now consists of six rows and two columns.

11. Continue adding columns and entering data as shown in Table 2.1 to create the final data table with six rows and five columns.

Changing the Data Table Name

1. Double-click on the data table name (Untitled) in the Table Panel.

2. Type the new name (Blood Pressure).

Figure 2.6 Changing the Data Table Name

Double-click here, and then type the new name.

Working with Data Tables

This section contains the following information:

- "Editing Data" on page 46

- "Selecting, Deselecting, and Finding Values" on page 47

- "Viewing or Changing Column Information" on page 51

- "Filtering Data" on page 54

- "Calculating Values with Formulas" on page 52

Editing Data

You can enter or change data, either a few cells at a time or for an entire column. This section contains the following information:

- "Changing Values" on page 46
- "Recoding Values" on page 46
- "Creating Patterned Data" on page 47

The examples in this section use the Companies.jmp sample data table. Before you continue, open this data table.

Changing Values

To change a value, select a cell and type the change. You can also double-click a cell to edit it.

Note: Double-clicking in a cell is not the same as selecting a cell. A single click selects a cell. You can select more than one cell at the same time, and you can perform certain actions on selected cells. Double-clicking only lets you edit a cell. For more information about selecting rows, columns, and cells, see "Selecting, Deselecting, and Finding Values" on page 47.

Recoding Values

Use the recoding tool to change all of the values in a column at once. For example, suppose you are interested in comparing the sales of computer and pharmaceutical companies. Your current company labels are Computer and Pharmaceutical. You want to change them to Technical and Drug. Going through all 32 rows of data and changing all the values would be tedious, inefficient, and error-prone, especially if you had many more rows of data. Recode is a better option.

1. Select the Type column by clicking once on the column heading.
2. Select **Cols > Recode**.
3. In the Recode window, enter the desired values in the **New Values** boxes. For this example, enter Technical in the Computer row, and Drug in the Pharmaceutical row.
4. Select the **In Place** option from the menu.

Figure 2.7 Recode Window

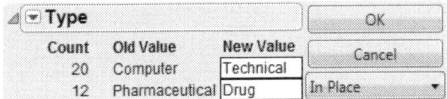

5. Click **OK**.

 All cells are updated automatically to the new values.

Creating Patterned Data

Use the Fill options to populate a column with patterned data. The Fill options are especially useful if your data table is large, and typing in the values for each row would be cumbersome.

Example: Filling a Column with the Pattern

1. Add a new column.

2. Enter 1 in the first cell, 2 in the second cell, and 3 in the third cell.

3. Select the three cells, and right-click anywhere in the selected cells to see a menu.

4. Select **Fill > Repeat sequence to end of table**.

 The rest of the column is filled with the sequence (1, 2, 3, 1, 2, 3, ...).

To continue a pattern instead of repeating it (1, 2, 3, 4, 5, 6, ...), select **Continue sequence to end of table**. This command can also be used to generate patterns like (1, 1, 1, 2, 2, 2, 3, 3, 3, ...).

The Fill options can recognize simple arithmetic and geometric sequences. For character data, the Fill options only repeat the values.

Selecting, Deselecting, and Finding Values

You can select rows, columns, or cells within a data table. For example, to create a subset of an existing data table, you must first select the parts of the table that you want to subset. Also, selecting rows can make data points stand out on a graph. Select rows and columns manually by clicking, or select rows that meet certain search criteria. This section contains the following information:

- "Selecting and Deselecting Rows" on page 47
- "Selecting and Deselecting Columns" on page 48
- "Selecting and Deselecting Cells" on page 49
- "Searching for Values" on page 49

Selecting and Deselecting Rows

Table 2.2 Selecting and Deselecting Rows

Task	Action
Select rows one at a time	Click on the row number.

Table 2.2 Selecting and Deselecting Rows

Task	Action
Select multiple adjacent rows	Click and drag on the row numbers. or Select the beginning row, and then hold down the SHIFT key and click the last row number.
Select multiple non-adjacent rows	Select the first row, and then hold down the CTRL key and click the other row numbers.
Deselect rows one at a time	Hold down the CTRL key and click the row numbers.
Deselect all rows	Click in the lower-triangular space in the top left corner of the table. See Figure 2.8.

Figure 2.8 Deselecting Rows

To deselect all rows at
once, click here.

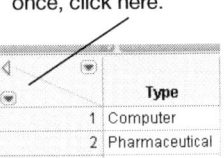

Selecting and Deselecting Columns

Table 2.3 Selecting and Deselecting Columns

Task	Action
Select columns one at a time	Click the column heading.
Select multiple adjacent columns	Click and drag across the column headings. or Select the beginning column, and then hold down the SHIFT key and click the last header.
Select multiple non-adjacent columns	Select the first column, and then hold down the CTRL key and click the other column headings.
Deselect columns one at a time	Hold down the CTRL key and click the column heading.

Table 2.3 Selecting and Deselecting Columns

Task	Action
Deselect all columns	Click in the upper-triangular space in the top left corner of the table. See Figure 2.9.

Figure 2.9 Deselecting Columns

To deselect all columns
at once, click here.

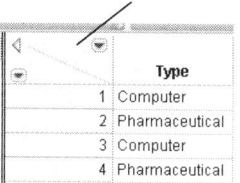

Selecting and Deselecting Cells

Table 2.4 Selecting and Deselecting Cells

Task	Action
Select cells one at a time	Click each cell individually.
Select multiple adjacent cells	Click and drag across the cells.
	or
	Select the beginning cell, and then hold down the SHIFT key and click the last cell.
Select multiple non-adjacent cells	Select the first cell, and then hold down the CTRL key and click the other cells.
Deselect all cells	Click in the upper and lower triangular spaces in the top left corner of the table.

Searching for Values

In a data table that has thousands or tens of thousands of rows, it can be difficult to locate a particular cell by scrolling through the table. If you are looking for specific information, use the Search feature to find it. If data is found that matches the search criteria, the cell is selected and the data grid scrolls to show it in the window. For example, the Companies.jmp data table contains information about a company that has total sales of $11,899. Use the Search feature to find that cell.

Example: Searching for a Value

1. Select **Edit > Search > Find** to launch the Search window.

2. In the **Find what** box, enter 11899.

3. Click **Find**. JMP finds the first cell that has 11,899 in it, and selects it.

If multiple cells meet the search criteria, click **Find** again to find the next cell that matches the search term.

You can also search for multiple rows at once, with each row matching some criteria.

Example: Select All of the Rows That Correspond to Medium-sized Companies

1. Select **Rows > Row Selection > Select Where** to open the **Select rows** window.

2. In the column list box on the left, select Size Co.

3. In the text box on the right, enter medium.

4. Click **OK**.

Figure 2.10 Select Rows Window

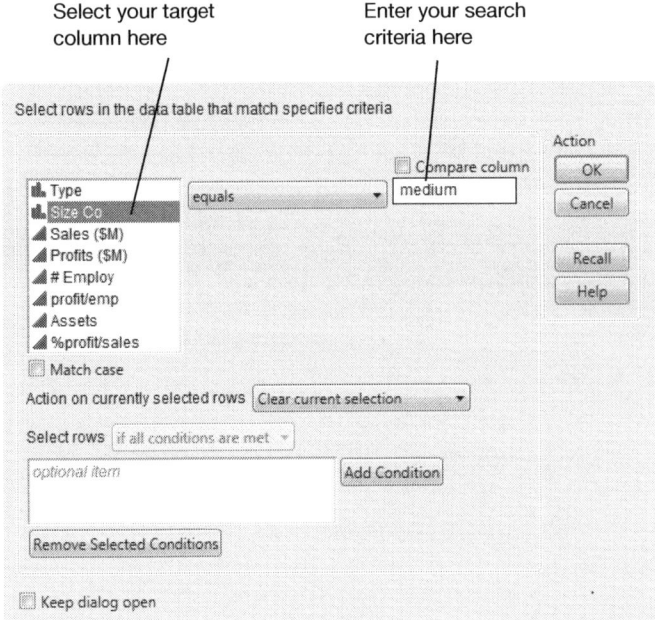

JMP selects all of the rows that have Size Co equal to medium. There are seven.

Viewing or Changing Column Information

Information about a column is not limited to the data in the column. Data type, modeling type, format, and formulas can also be set.

To view or change column characteristics, double-click on the column heading. Or, right-click on the column heading and select **Column Info**. The Column Info window appears.

Figure 2.11 Column Info Window

Table 2.5 Column Information

Option	Description
Column Name	Enter or change the column name. No two columns can have the same column name.
Data Type	Select one of the following data types:
	• **Numeric** specifies the column values as numbers.
	• **Character** specifies the column values as non-numeric, such as letters or symbols.
	• **Row State** specifies the column values as row states. This is an advanced topic. See *Using JMP*.

Table 2.5 Column Information *(Continued)*

Option	Description
Modeling Type	Modeling types define how values are used in analyses. Select one of the following modeling types: • **Continuous** values are numeric only. • **Ordinal** values are either numeric or character, and are ordered categories. • **Nominal** values are either numeric or character, but not ordered.
Format	Select a format for numeric values. This option is not available for character data. Here are a few of the most common formats: • **Best** lets JMP choose the best display format. • **Fixed Dec** specifies the number of decimal places that appear. • **Date** specifies the syntax for date values. • **Time** specifies the syntax for time values. • **Currency** specifies the type of currency and decimal points that are used for currency values.
Column Properties	Set special column properties such as formulas, notes, and value orders. See *Using JMP*.
Lock	Lock a column, so that the values in the column cannot be changed.

Calculating Values with Formulas

Use the Formula Editor to create columns that contain calculated values.

Scenario

The sample data table On-Time Arrivals.jmp reflects the percent of on-time arrivals for several airlines. The data was collected for March, June, and August of 1999.

Creating the Formula

Suppose that you want to create a new column containing the average on-time percentage for each airline.

1. Add a new column.

2. Right-click on the column heading of the new column and select **Formula**. The Formula Editor window appears.

Figure 2.12 Formula Editor

Create the formula for the average on-time percentage of each airline:

3. From the Table Columns list, select March 1999.

4. Click the ✚ button on the keypad.

5. Select June 1999, followed by another ✚ sign.

6. Select August 1999.

Figure 2.13 Sum of the Months

March 1999 + June 1999 + August 1999

Notice that only August 1999 is selected (has the red box around it).

7. Click on the box surrounding the entire formula.

Figure 2.14 Entire Formula Selected

March 1999 + June 1999 + August 1999

8. Click the button.

9. Type a 3 in the denominator box, and then click outside of the formula in any of the white space.

Figure 2.15 Completed Formula

$$\frac{March\ 1999 + June\ 1999 + August\ 1999}{3}$$

10. Click **OK**

The new column contains the averages.

The Formula Editor has many built-in arithmetic and statistical functions. For example, another way to calculate the average on-time arrival percentage is to use the **Mean()** function in the Statistical functions list. For details about all of the Formula Editor functions, see *Using JMP*.

Filtering Data

Use the **Data Filter** to interactively select complex subsets of data, hide these subsets in plots, or exclude them from analyses. For example, look at profit per employee for computer and pharmaceutical companies.

1. Open the Companies.jmp sample data table.

2. Select **Analyze** > **Distribution**.

3. Select profit/emp and click **Y, Columns**.

4. Click **OK**.

5. From the red triangle menu for profit/emp, select **Display Options** > **Horizontal Layout**.

Figure 2.16 Distribution of profit/emp

6. Turn on Automatic Recalc by selecting **Script** > **Automatic Recalc** from the red triangle menu for **Distributions**.

When this option is on, every change that you make (for example, hiding or excluding points) causes your report window to automatically update itself.

7. Select **Rows** > **Data Filter**.

8. Select Type and click **Add**.

9. Clear **Select** and select **Include**.

10. To filter out the Pharmaceutical companies from the Distribution results, and include only the Computer companies, click the **Computer** box in the Data Filter window.

The distribution results update to only include Computer companies.

Figure 2.17 Filter for Computer Companies

Click the Computer box to include only computer companies in the Distribution results.

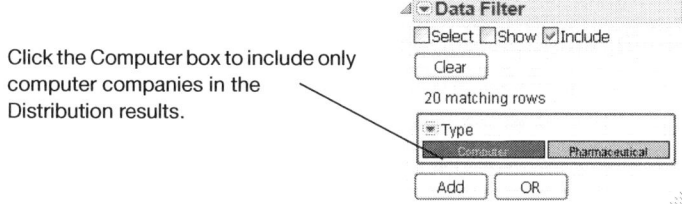

The graph and the statistics report automatically reflect only the rows that are selected.

Conversely, to change the Distribution results to include only the Pharmaceutical companies, click the **Pharmaceutical** button on the Data Filter window.

Managing Data

The commands on the **Tables** menu (and Tabulate on the **Analyze** menu) summarize and manipulate data tables into the format that you need for graphing and analyzing. This section describes five of these commands:

Summary Creates a table that contains summary statistics that describe your data.

Tabulate Provides a drag and drop workspace to create summary statistics.

Subset Creates a table that contains a subset of your data.

Join Joins the data from two data tables into one new data table.

Sort Sorts your data by one or more columns.

For complete details about these and the other Tables menu commands, see *Using JMP*.

Requesting Summary Statistics

Summary statistics, such as sums and means, can instantly provide useful information about your data. For example, if you look at the annual profit of each company out of thirty-two companies, it's difficult to compare the profits of small, medium, and large companies. A summary shows that information immediately.

Create summary tables by using either the **Summary** or **Tabulate** commands. The **Summary** command creates a new data table. As with any data table, you can perform analyses and create graphs from the summary table. The **Tabulate** command creates a report window with a table of summary data. You can also create a table from the Tabulate report.

Summary

A summary table contains statistics for each level of a grouping variable. For example, look at the financial data for computer and pharmaceutical companies. Suppose that you want to calculate the mean of sales and the mean of profits, for each combination of company type and size.

1. Open the Companies.jmp sample data table.
2. Select **Tables > Summary**.
3. Select Type and Size Co and click **Group**.
4. Select Sales ($M) and Profits ($M) and click **Statistics > Mean**.

Figure 2.18 Completed Summary Window

5. Click **OK**.

JMP calculates the mean of Sales ($M) and the mean of Profit ($M) for each combination of Type and Size Co.

Figure 2.19 Summary Table

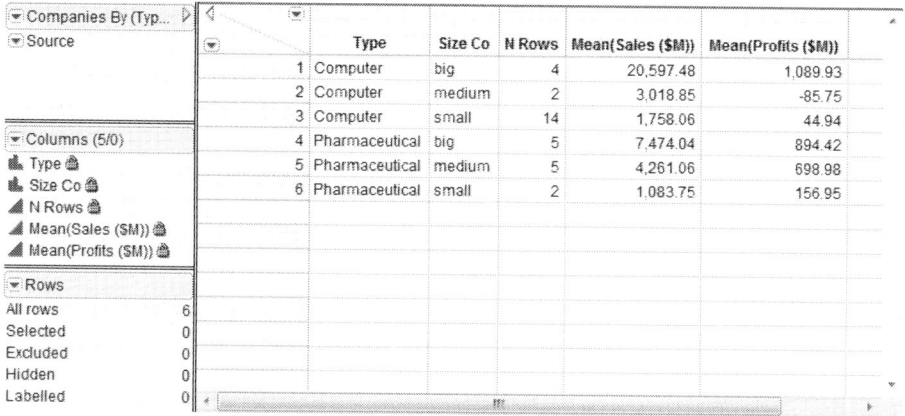

The summary table contains the following:

- There are columns for each grouping variable (in this example, Type, and Size Co).

- The N Rows column shows the number of rows from the original table that correspond to each combination of grouping variables. For example, the original data table contains 14 rows corresponding to small computer companies.

- There is a column for each summary statistic requested. In this example, there is a column for the mean of Sales ($M) and a column for the mean of Profits ($M).

The summary table is linked to the source table. Selecting a row in the summary table also selects the corresponding rows in the source table.

Tabulate

Use the Tabulate command to drag columns into a workspace, creating summary statistics for each combination of grouping variables. This example shows you how to use Tabulate to create the same summary information that you just created using Summary.

1. Open the Companies.jmp sample data table.

2. Select **Analyze** > **Tabulate**.

Figure 2.20 Tabulate Workspace

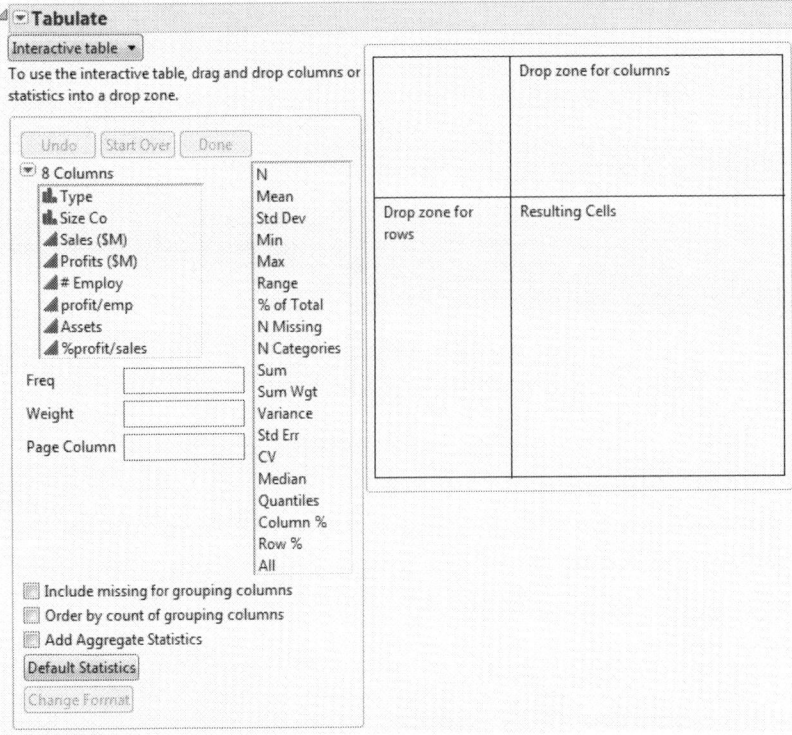

3. Select both Type and Size Co.

4. Drag and drop them into the **Drop zone for rows.**

Figure 2.21 Dragging Columns to the Row Zone

5. Right-click on a heading and select **Nest Grouping Columns.**

 The initial tabulation shows the number of rows per group.

Figure 2.22 Initial Tabulation

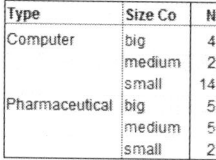

Type	Size Co	N
Computer	big	4
	medium	2
	small	14
Pharmaceutical	big	5
	medium	5
	small	2

6. Select both Sales ($M) and Profits ($M), and drag and drop them over the **N** in the table.

Figure 2.23 Adding Sales and Profit

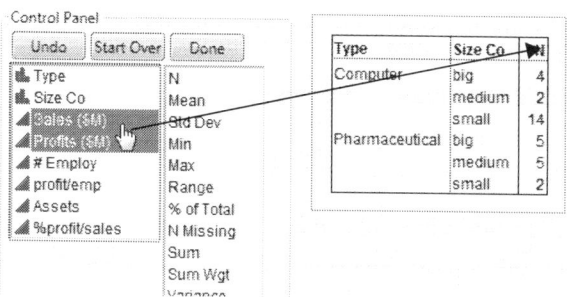

The tabulation now shows the sum of Sales ($M) and the sum of Profits ($M) per group.

Figure 2.24 Tabulation of Sums

Type	Size Co	Sales ($M) Sum	Profits ($M) Sum
Computer	big	82389.9	4359.7
	medium	6037.7	-171.5
	small	24612.8	629.1
Pharmaceutical	big	37370.2	4472.1
	medium	21305.3	3494.9
	small	2167.5	313.9

7. The final step is to change the sums to means. Right-click on **Sum** (either of them) and select **Statistics** > **Mean**.

Figure 2.25 Final Tabulation

Type	Size Co	Sales ($M) Sum	Mean	Profits ($M) Sum	Mean
Computer	big	82389.9	20597.48	4359.7	1089.9
	medium	6037.7	3018.85	-171.5	-85.75
	small	24612.8	1758.06	629.1	44.94
Pharmaceutical	big	37370.2	7474.04	4472.1	894.42
	medium	21305.3	4261.06	3494.9	698.98
	small	2167.5	1083.75	313.9	156.95

The means are the same as those obtained using the Summary command. Compare Figure 2.25 to Figure 2.19.

Creating Subsets

If you want to look closely at only part of your data table, you can create a subset. For example, suppose that you have already compared the sales and profits of big, medium, and small computer and pharmaceutical companies. Now you want to look at the sales and profits of only the medium-sized companies.

Creating a subset is a two-step process. First select the target data, and then extract the data into a new table.

Subsetting with the Subset Command

1. Open the Companies.jmp sample data table.

Selecting the Rows and Columns That You Want to Subset

2. Select **Rows > Row Selection > Select Where**.

3. Select Size Co in the column list box on the left.

4. Enter medium in the text enter box.

5. Click **OK**.

6. Select the Type, Sales ($M), and Profits ($M) columns.

Creating the Subset Table

7. Select **Tables > Subset** to launch the Subset window.

Figure 2.26 Subset Window

8. Select **Selected columns** to subset only the columns that you selected. You can also customize your subset table further by selecting additional options.

9. Click **OK**.

The resulting subset data table has seven rows and three columns. For complete details about the Subset command, see *Using JMP*.

Subsetting with the Distribution Platform

Another way to create subsets uses the connection between platform results and data tables.

Example: Creating a Subset Using the Distribution Command

1. Open the Companies.jmp sample data table.

2. Select **Analyze > Distribution**.

3. Select Type and click **Y, Columns**.

4. Click **OK**.

5. Double-click on the histogram bar that represents Computer to create a subset table of the Computer companies.

Caution: This method creates a *linked* subset table. This means if you make any changes to the data in the subset table, the corresponding value changes in the source table.

Joining Data Tables

Use the Join option to combine information from multiple data tables into a single data table. For example, suppose that you have a data table containing results from an experiment on popcorn yields. In another data table, you have the results of a second experiment on popcorn yields. To compare the two experiments or to analyze the trials using both sets of results, you need to have the data in the same table. Also, the experimental data was not entered into the data tables in the same order. One of the columns has a different name, and the second experiment is incomplete. This means that you cannot copy and paste from one table into another.

Example: Joining Two Data Tables

1. Open the Trial1.jmp and Little.jmp data tables.
2. Click on Trial1.jmp to make it the active data table.
3. Select **Tables > Join.**
4. In the **Join 'Trial1' With** box, select Little.
5. From the **Matching Specification** menu, select **By Matching Columns.**
6. In the **Source Columns** boxes, select popcorn in both boxes, and then click **Match.**
7. In the same way, match batch and oil amt to oil in both boxes.

 Your matching columns do not have to have the same name.
8. Select **Include non-matches** for both tables.

 Since one experiment is partial, you want to include all rows, including any with missing data.
9. To avoid duplicate columns, select the **Select columns for joined table** option.
10. From Trial1, select all four columns and click **Select.**
11. From Little, select only yield and click **Select.**

Figure 2.27 Completed Join Window

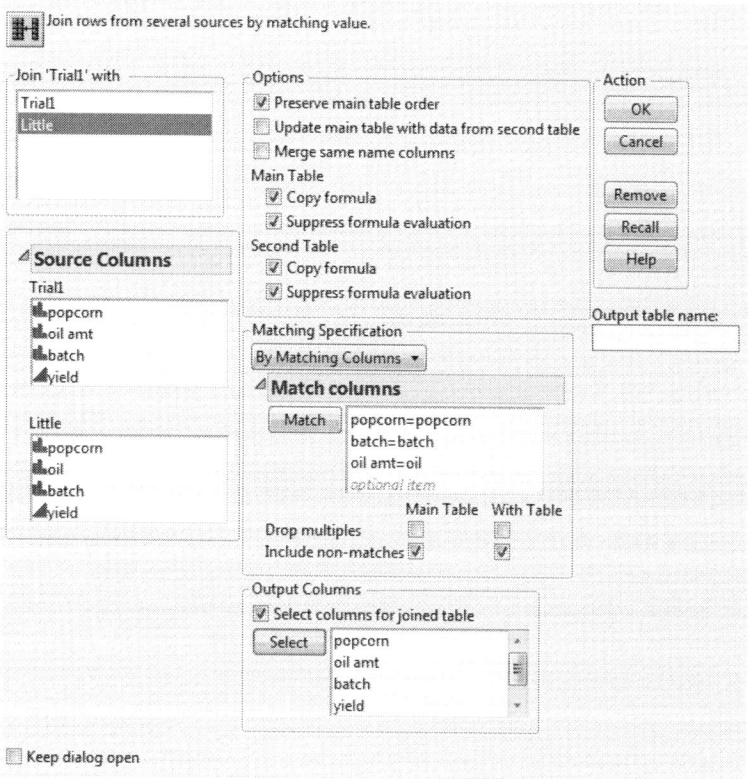

12. Click **OK**.

Figure 2.28 Joined Table

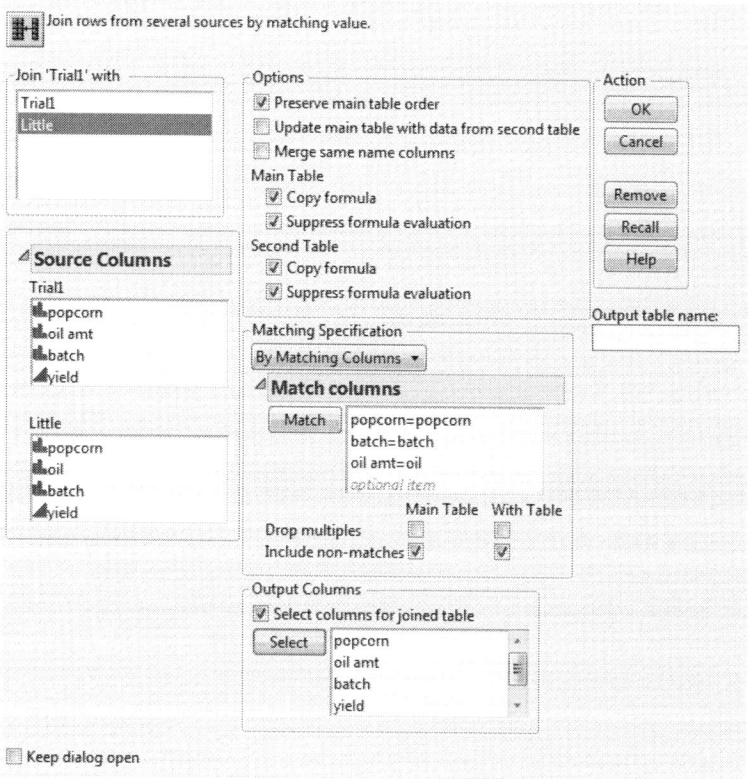

Sorting Tables

Use the Sort command to sort a data table by one or more columns in the data table. For example, look at financial data for computer and pharmaceutical companies. Suppose that you want to sort the data table by Type, then by Profits ($M). Also, you want Profits ($M) to be in descending order within each Type.

1. Open the Companies.jmp sample data table.

2. Select **Tables** > **Sort**.

3. Select Type and click **By** to assign Type as a sorting variable.

4. Select Profits ($M) and click **By**.

At this point, both variables are set to be sorted in ascending order. See the ascending icon next to the variables in Figure 2.29.

Figure 2.29 Sort Ascending Icon

5. To change Profits ($M) to sort in descending order, select Profits ($M) and click the descending button.

Figure 2.30 Change Profits to Descending

The icon next to Profits ($M) changes to descending.

6. Select the **Replace Table** check box.

 When selected, the **Replace Table** option tells JMP to sort the original data table instead of creating a new table with the sorted values. This option is not available if there are any open report windows created from the original data table. Sorting a data table with open report windows might change how some of the data is displayed in the report window, especially in graphs.

7. Click **OK**.

The data table is now sorted by type alphabetically, and by descending profit totals within type.

Chapter **3**

Visualizing Your Data
Graphing Data

Visualizing your data is an important first step. The graphs described in this chapter help you discover important details about your data. For example, histograms show you the shape and range of your data, and help you find unusual data points.

This chapter presents several of the most common graphs and plots that enable you to visualize and explore data in JMP. This chapter is an introduction to some of JMP's graphical tools and platforms. Use JMP to visualize the distribution of single variables, or the relationships among multiple variables.

Figure 3.1 Visualizing Data with JMP

Contents

Looking at Single Variables

Single-variable graphs, or *univariate* graphs, let you look closely at one variable at a time. When you begin to look at your data, it's important to learn about each variable before looking at how the variables interact with each other. Univariate graphs let you visualize each variable individually.

This section covers two graphs that show the distribution of a single variable:

- "Histograms" on page 69, for continuous variables
- "Bar Charts" on page 71, for categorical variables

Use the Distribution platform to create both of these graphs. Distribution produces a graphical description and descriptive statistics for each variable.

Histograms

The histogram is one of the most useful graphical tools for understanding the distribution of a continuous variable. Use a histogram to find the following in your data:

- the average value and variation
- extreme values

Figure 3.2 Example of a Histogram

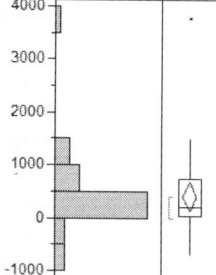

Scenario

This example uses the Companies.jmp data table, which contains data on profits for a group of companies.

A financial analyst wants to explore the following questions:

- Generally, how much profit does each company earn?
- What is the average profit?

- Are there any companies that earn either extremely high or extremely low profits compared to the other companies?

To answer these questions, use a histogram of Profits ($M).

Creating the Histogram

1. Open the Companies.jmp sample data table.
2. Select **Analyze** > **Distribution**.
3. Select Profits ($M) and click **Y, Columns**.

Figure 3.3 Distribution Window for Profits ($M)

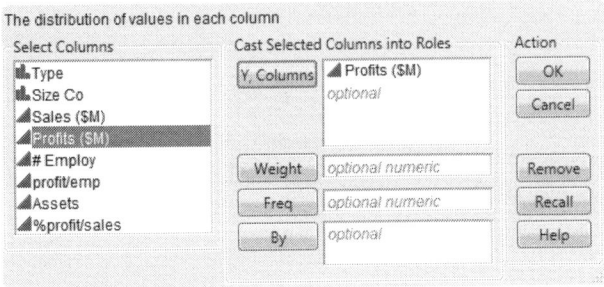

4. Click **OK**.

Figure 3.4 Histogram of Profits ($M)

Interpreting the Histogram

The histogram provides these answers:

- Most companies' profits are between $-1000 and $1500.

 All the bars except for one are located in this range. Also, more companies' profits range from $0 to $500 than any other range. The bar representing that range is much longer than the others.

- The average profit is a little less than $500.

 The middle of the diamond in the box plot indicates the mean value. In this case, the mean is slightly lower than the $500 mark.

- One company has significantly higher profits than the others, and might be an *outlier*. An outlier is a data point that is separated from the general pattern of the other data points.

 This outlier is represented by a single, very short bar at the top of the histogram. The bar is small and represents a small group (in this case, a single company), and it is widely separated from the rest of the histogram bars.

In addition to the histogram, this report includes the following:

- The box plot, which is another graphical summary of the data. For detailed information about the box plot, see the *Essential Graphing* book.

- **Quantiles** and **Summary Statistics** reports. These reports are discussed in "Analyzing Distributions" on page 107 in the "Analyzing Your Data" chapter.

Interacting with the Histogram

Data tables and reports are all connected in JMP. Click on a histogram bar to select the corresponding rows in the data table.

Bar Charts

Use a bar chart to visualize the distribution of a categorical variable. A bar chart looks similar to a histogram, since they both have bars that correspond to the levels of a variable. A bar chart shows a bar for every level of the variable, whereas the histogram shows a range of values for the variable.

Figure 3.5 Example of a Bar Chart

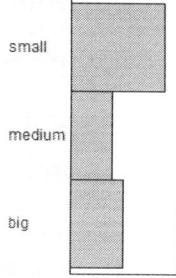

Scenario

This example uses the Companies.jmp data table, which contains data on the size and type of a group of companies.

A financial analyst wants to explore the following questions:

- What is the most common type of company?
- What is the most common size for a company?

To answer these questions, use bar charts of Type and Size Co.

Creating the Bar Chart

1. Open the Companies.jmp sample data table.
2. Select **Analyze** > **Distribution**.
3. Select Type and Size Co and click **Y, Columns**.
4. Click **OK**.

Figure 3.6 Bar Charts of Type and Size Co

Interpreting the Bar Charts

The bar charts provides these answers:

- There are more computer companies than pharmaceutical companies.

 The bar that represents computer companies is larger than the bar that represents pharmaceutical companies.

- The most common company size is small.

 The bar that represents small companies is larger than the bars that represent medium and big companies.

The additional summary output gives detailed frequencies. This report is discussed in "Distributions of Categorical Variables" on page 110 in the "Analyzing Your Data" chapter.

Interacting with the Bar Charts

As is the case with histograms, click on individual bars to highlight rows of the data table. If more than one graph is created, clicking on a bar in one bar chart highlights the corresponding bar or bars in the other bar chart.

For example, suppose that you want to see the distribution of company size for the pharmaceutical companies. Click on the Pharmaceutical bar in the Type bar chart, and the pharmaceutical companies are highlighted on the Size Co bar chart. Figure 3.7 shows that although most companies in this data table are small, most of the pharmaceutical companies are medium or big.

Also, the corresponding rows in the data table are selected.

Figure 3.7 Clicking Bars

Click on this bar to select the corresponding data in the other chart.

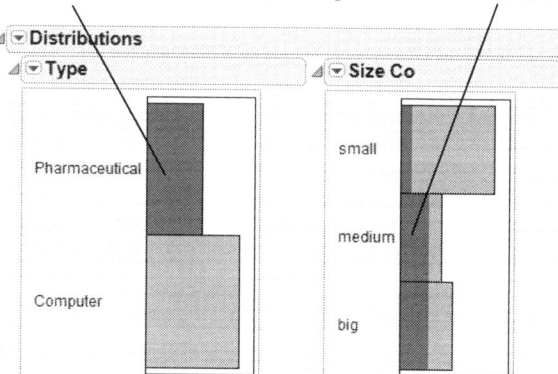

Comparing Multiple Variables

Use multiple-variable graphs to visualize the relationships and patterns between two or more variables. This section covers the following graphs:

Table 3.1 Multiple-Variable Graphs

"Scatterplots" on page 74	Use scatterplots to compare two continuous variables.
"Scatterplot Matrix" on page 78	Use scatterplot matrices to compare several pairs of continuous variables.

Table 3.1 Multiple-Variable Graphs

"Side-by-Side Box Plots" on page 80	Use side-by-side box plots to compare one continuous and one categorical variable.
"Overlay Plots" on page 83	Use overlay plots to compare one or more variables on the Y-axis to another variable on the X-axis. Overlay plots are especially useful if the X variable is a time variable, because you can compare how two or more variables change across time.
"Variability Chart" on page 86	Use variability charts to compare one continuous Y variable to one or more categorical X variables. Variability charts show differences in means and variability across several categorical X variables.
"Graph Builder" on page 89	Use Graph Builder to create and change graphs interactively.
"Bubble Plots" on page 93	Bubble plots are specialized scatterplots that use color and bubble sizes to represent up to five variables at once. If one of your variables is a time variable, you can animate the plot to see your other variables change through time.

Scatterplots

The scatterplot is the simplest of all the multiple-variable graphs. Use scatterplots to determine the relationship between two continuous variables and to discover whether two continuous variables are *correlated*. Correlation indicates how closely two variables are related. When you have two variables that are highly correlated, one might influence the other. Or, both might be influenced by other variables in a similar way.

Figure 3.8 Example of a Scatterplot

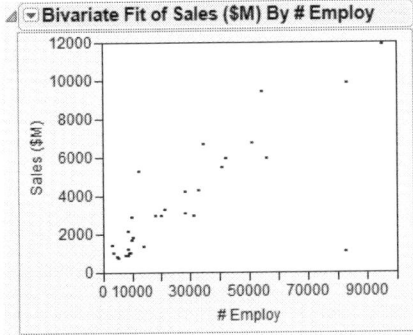

Scenario

This example uses the Companies.jmp data table, which contains sales figures and the number of employees of a group of companies.

A financial analyst wants to explore the following questions:

• What is the relationship between sales and the number of employees?

• Does the amount of sales increase with the number of employees?

• Can you predict average sales from the number of employees?

To answer these questions, use a scatterplot of Sales ($M) versus # Employ.

Creating the Scatterplot

1. Open the Companies.jmp sample data table.

2. Select **Analyze > Fit Y by X**.

3. Select Sales ($M) and **Y, Response**.

4. Select # Employ and **X, Factor**.

Figure 3.9 Fit Y by X Window

5. Click **OK**.

Figure 3.10 Scatterplot of Sales ($M) versus # Employ

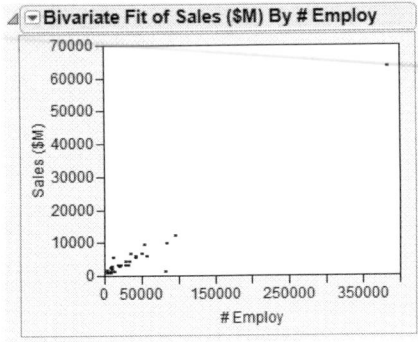

Interpreting the Scatterplot

One company has a large number of employees and high sales, represented by the single point at the top right of the plot. The distance between this data point and all the rest makes it difficult to visualize the relationship between the rest of the companies. Remove the point from the plot and recreate the plot by following these steps:

1. Click on the point to select it.

2. Select **Rows > Hide and Exclude**. The data point is hidden and no longer included in calculations.

Note: The difference between hiding and excluding is important. Hiding a point removes it from any graphs but statistical calculations continue to use the point. Excluding a point removes it from any statistical calculations but does not remove it from graphs. When you both hide and exclude a point, you remove it from all calculations and from all graphs.

3. To recreate the plot without the outlier, select **Script > Redo Analysis** from the red triangle menu for Bivariate. You can close the original report window.

Figure 3.11 Scatterplot with the Outlier Removed

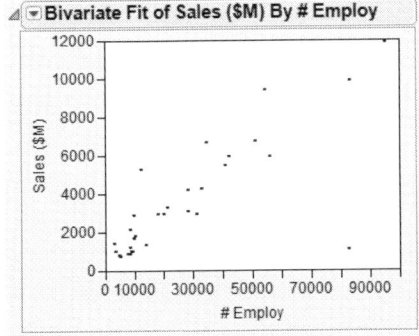

The updated scatterplot provides these answers:

- There is a relationship between the sales and the number of employees.

 The data points have a discernible pattern. They are not scattered randomly throughout the graph. You could draw a diagonal line that would be near most of the data points.

- Sales do increase with the number of employees, and the relationship is linear.

 If you drew that diagonal line, it would slope from bottom left to top right. This slope shows that as the number of employees increases (left to right on the bottom axis), sales also increases (bottom to top on the left axis). A straight line would be near most of the data points, indicating a linear relationship. If you would have to curve your line to be near the data points, there would still be a relationship (because of the pattern of the points). However, that relationship would not be linear.

- You can predict average sales from the number of employees.

 The scatterplot shows that sales generally increase as the number of employees does. You could predict the sales for a company if you knew only the number of employees of that company. Your prediction would be on that imaginary line. It would not be exact, but it would approximate the real sales.

Interacting with the Scatterplot

As with other JMP graphics, the scatterplot is interactive. Hover over the point in the bottom right corner with the mouse to reveal the row number and the x and y values.

Figure 3.12 Hover Over a Point

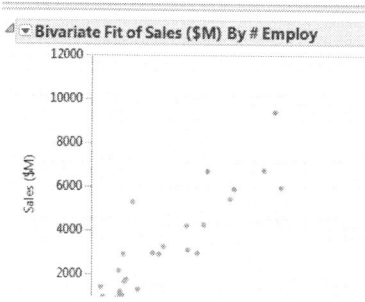

Click on a point to highlight the corresponding row in the data table. Select multiple points by doing one of the following:

- Click and drag with the mouse around the points. This selects points in a rectangular area.
- Select the lasso tool, and then click and drag around multiple points. The lasso tool selects an irregularly shaped area.

Scatterplot Matrix

A scatterplot matrix is a collection of scatterplots organized into a grid (or matrix). Each scatterplot shows the relationship between a pair of variables.

Figure 3.13 Example of a Scatterplot Matrix

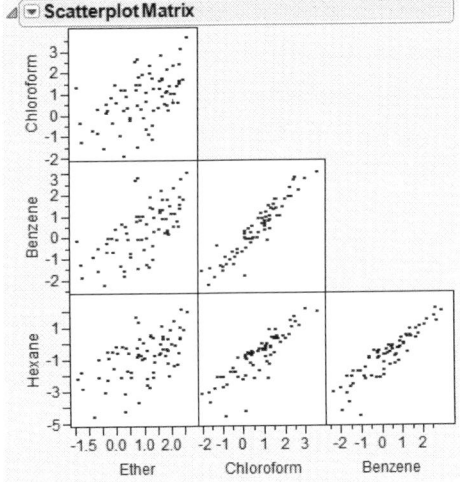

Scenario

This example uses the Solubility.jmp data table, which contains data for solubility measurements for 72 different solutes.

A lab technician wants to explore the following questions:

- Is there a relationship between any pair of chemicals? (There are six possible pairs.)
- Which pair has the strongest relationship?

To answer these questions, use a scatterplot matrix of the four solvents.

Creating the Scatterplot Matrix

1. Open the Solubility.jmp sample data table.
2. Select **Graph** > **Scatterplot Matrix**.
3. Select Ether, Chloroform, Benzene, and Hexane, and click **Y, Columns**.

Figure 3.14 Scatterplot Matrix Window

4. Click **OK**.

Figure 3.15 Scatterplot Matrix

Interpreting the Scatterplot Matrix

The scatterplot matrix provides these answers:

* All six pairs of variables are positively correlated.

 As one variable increases, the other variable increases too.

* The strongest relationship appears to be between Benzene and Chloroform.

 The data points in the scatterplot for Benzene and Chloroform are the most tightly clustered along an imaginary line.

Interacting with the Scatterplot Matrix

If you select a point in one scatterplot, it is selected in all the other scatterplots.

For example, if you select a point in the Benzene versus Chloroform scatterplot, the same point is selected in the other five plots.

Figure 3.16 Selected Points

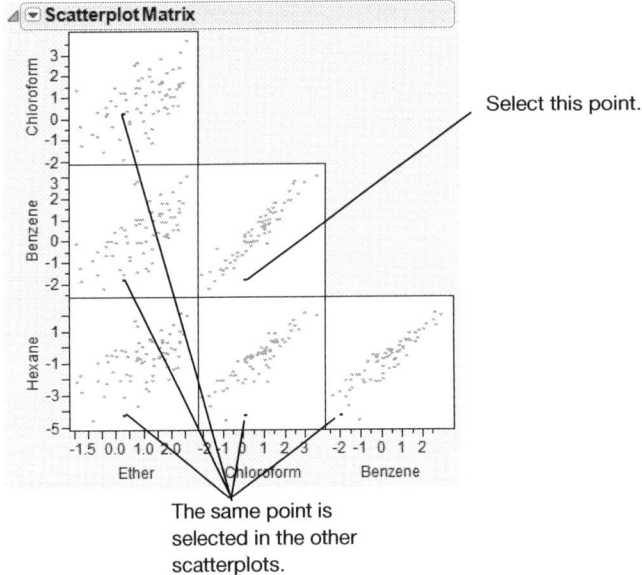

Select this point.

The same point is selected in the other scatterplots.

Side-by-Side Box Plots

Side-by-side box plots show the following:

- the relationship between one continuous variable and one categorical variable
- differences in the continuous variable across levels of the categorical variable

Figure 3.17 Example of Side-by-Side Box Plots

Scenario

This example uses the Analgesics.jmp data table, which contains data on pain measurements taken on patients using three different drugs.

A researcher wants to explore the following questions:

- Are there differences in the average amount of pain control among the drugs?

- Does the *variability* in the pain control given by each drug differ? A drug with high variability would not be as reliable as a drug with low variability.

To answer these questions, use a side-by-side box plot for the pain levels and the drug categories.

Creating the Side-by-Side Box Plots

1. Open the Analgesics.jmp data table.
2. Select **Analyze** > **Fit Y by X**.
3. Select pain and click **Y, Response**.
4. Select drug and click **X, Factor**.

Figure 3.18 Fit Y by X Window

5. Click **OK**.

6. From the red triangle menu, select **Display Options > Box Plots**.

Figure 3.19 Side-by-Side Box Plots

Interpreting the Side-by-Side Box Plots

Box plots are designed according to the following principles:

- The line through the box represents the median.
- The middle half of the data is within the box.
- The majority of the data falls between the ends of the whiskers.
- A data point outside the whiskers might be an outlier.

The box plots in Figure 3.19 show these answers:

- There is evidence to believe that patients on drug A feel less pain, since the box plot for drug A is lower on the pain scale than the others.
- Drug B appears to have higher variability than Drugs A and C, since the box plot is taller.

There is one point for drug C that is a lot lower than the other points for drug C. Hover over it with your mouse to see that it is row 26 of the data table. That point looks like it is more similar to the data in drug group A or B. The information in row 26 deserves investigation. There might have been a typographical error when the data was recorded.

Overlay Plots

Like scatterplots, overlay plots show the relationship between two or more variables. However, if one of the variables is a time variable, an overlay plot shows trends across time better than scatterplots do.

Figure 3.20 Example of an Overlay Plot

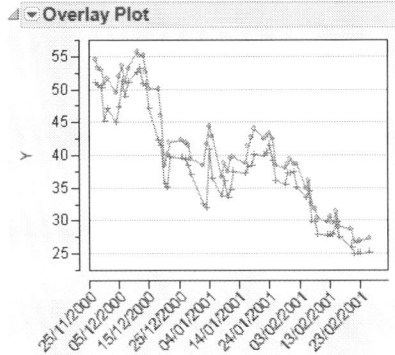

Note: To plot data over time, you can also use Graph Builder, bubble plots, control charts, and variability charts. For complete details on Graph Builder and bubble plots, see the *Essential Graphing* book. Refer to the *Quality and Process Methods* book for information about control charts and variability charts.

Scenario

This example uses the Stock Prices.jmp data table, which contains data on the price of a stock over a three-month period.

A potential investor wants to explore the following questions:

- Has the stock's closing price changed over the past three months?

 To answer this question, use an overlay plot of the stock's closing price over time.

- How do the stock's high and low prices relate to each other?

 To answer this question, use another overlay plot of the stock's high and low prices over time.

Create the first overlay plot to answer the first question, and then create a second overlay plot to answer the second question.

Creating the Overlay Plot of the Stock's Price over Time

1. Open the Stock Prices.jmp data table.
2. Select **Graph** > **Overlay Plot**.
3. Select Close and click **Y.**
4. Select Date and click **X.**

Figure 3.21 Overlay Plot Window

5. Click **OK.**

Figure 3.22 Overlay Plot of the Closing Price Over Time

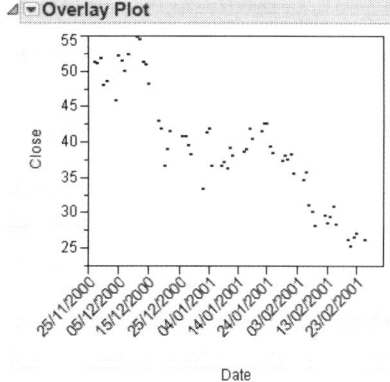

Interpreting and Interacting with the Overlay Plot

The overlay plot shows that the closing stock price has been decreasing over the last several months. To see the trend more clearly, connect the points and add grid lines.

1. From the red triangle menu, select **Connect Thru Missing**.

2. Double-click on the Y axis.

3. Select the **Major Gridline** check box.

4. Click **OK**.

Figure 3.23 Connected Points and Grid Lines

The potential investor can see that although the stock price has gone up and down over the past three months, the overall trend has been downward.

Creating the Overlay Plot of the Stock's High and Low Prices

Use an overlay plot to plot more than one Y variable. For example, suppose that you want to see both the high and the low prices on the same plot.

1. Follow the steps in "Creating the Overlay Plot of the Stock's Price over Time" on page 84, this time assigning both High and Low to the **Y** role.

2. Connect the points and add grid lines as shown in "Interpreting and Interacting with the Overlay Plot" on page 85.

Figure 3.24 Two Y Variables

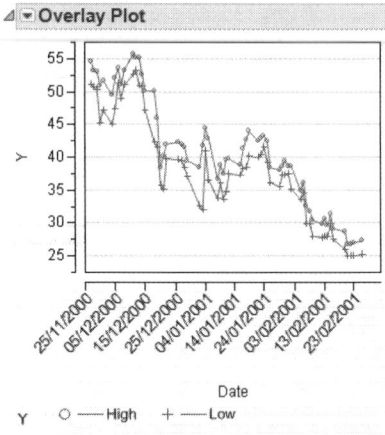

The legend at the bottom of the plot shows the colors and markers used for the High and Low variables in the graph. The overlay plot shows that the High price and Low price track each other very closely.

Answering the Questions

Both of the overlay plots answer the two questions asked at the beginning of this example.

- The first plot shows that the price of this stock has not remained the same, but has been decreasing.

- The second plot shows that the high and low prices of this stock are not very different from each other. The stock price does not vary wildly on any given day.

Variability Chart

In the graphs described so far, you specified only a single X variable. Use a variability chart to specify multiple X variables and see differences in means and variability across all of your variables at once.

Figure 3.25 Example of a Variability Chart

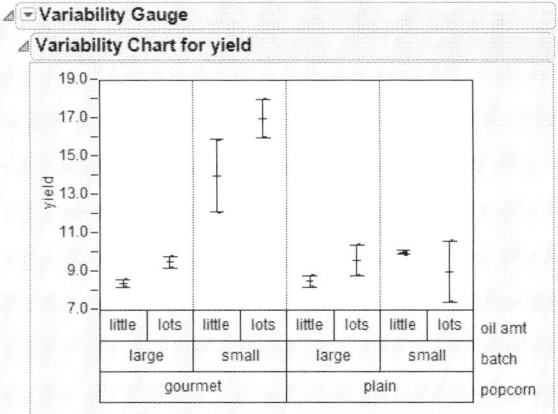

Scenario

This example uses the Popcorn.jmp data table with data from a popcorn maker. The yield (the volume of popcorn for a given measure of kernels) was measured for each combination of popcorn style, batch size, and amount of oil used.

The popcorn maker wants to explore the following question:

- Which combination of factors results in the highest popcorn yield?

To answer this question, use a variability chart of the yield versus the style, batch size, and oil amount.

Creating the Variability Chart

1. Open the Popcorn.jmp data table.
2. Select **Analyze > Quality and Process > Variability/Attribute Gauge Chart**.
3. Select yield and click **Y, Response**.
4. Select popcorn and click **X, Grouping**.
5. Select batch and click **X, Grouping**.
6. Select oil amt and click **X, Grouping**.

Note: The order in which you assign the variables to the **X, Grouping** role is important, because the order in this window determines their nesting order in the variability chart.

Figure 3.26 Variability Chart Window

7. Click **OK**.

The top chart is the variability chart, showing the yield broken down by each combination of the three variables. The bottom chart shows the standard deviation for each combination of the three variables. Since the bottom chart does not show the yield, hide it.

8. Deselect **Std Dev Chart** on the red triangle menu.

Figure 3.27 Results Window

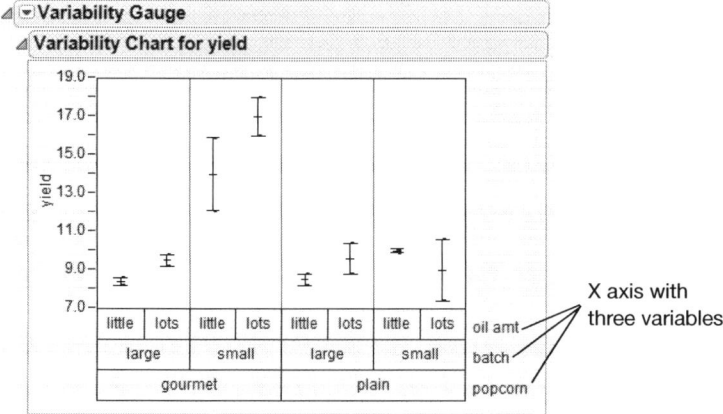

Interpreting the Variability Chart

The variability chart for yield indicates that small, gourmet batches produce the highest yield.

To be more specific, the popcorn maker might ask this additional question: Is the yield high because those batches are small, or because those batches are gourmet?

The variability chart shows the following:

- The yield from small, plain batches is low.

- The yield from large, gourmet batches is low.

Given this information, the popcorn maker can conclude that only the combination of small and gourmet at the same time results in batches with high yield. It would have been impossible to reach this conclusion with a chart that only allowed a single variable.

Graph Builder

Use Graph Builder to interactively create and modify graphs. So far, all of the graphs have been created by launching a platform and specifying variables. To create a different type of graph, you must launch a different platform. In Graph Builder, you can change the variables and change the graphs at any time.

Use Graph Builder to accomplish the following tasks:

- Change variables by dragging and dropping them in and out of the graph.

- Create a different type of graph with a few mouse clicks.

- Partition the graph horizontally or vertically.

Figure 3.28 Example of a Graph that was Created with Graph Builder

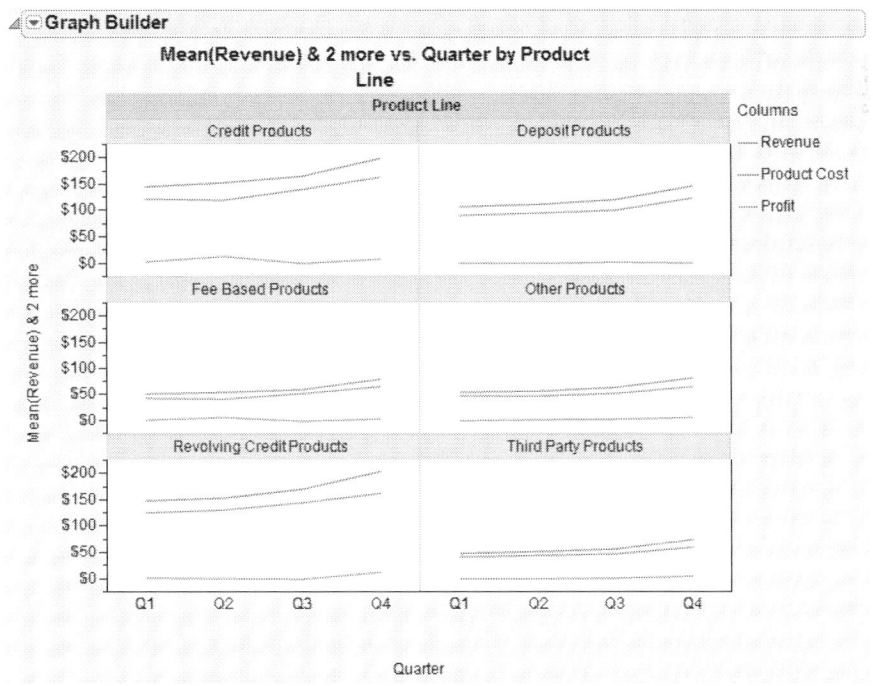

Note: Only some of the Graph Builder features are covered here. For complete details, see the *Essential Graphing* book.

Scenario

This example uses the Profit by Product.jmp data table, which contains profit data for multiple product lines.

A business analyst wants to explore the following question:

- How is the profitability different between product lines?

To answer this question, use a line plot that displays revenue, product cost, and profit data across different product lines.

Creating the Graph

1. Open the Profit by Product.jmp data table.

2. Select **Graph** > **Graph Builder**.

Figure 3.29 Graph Builder Workspace

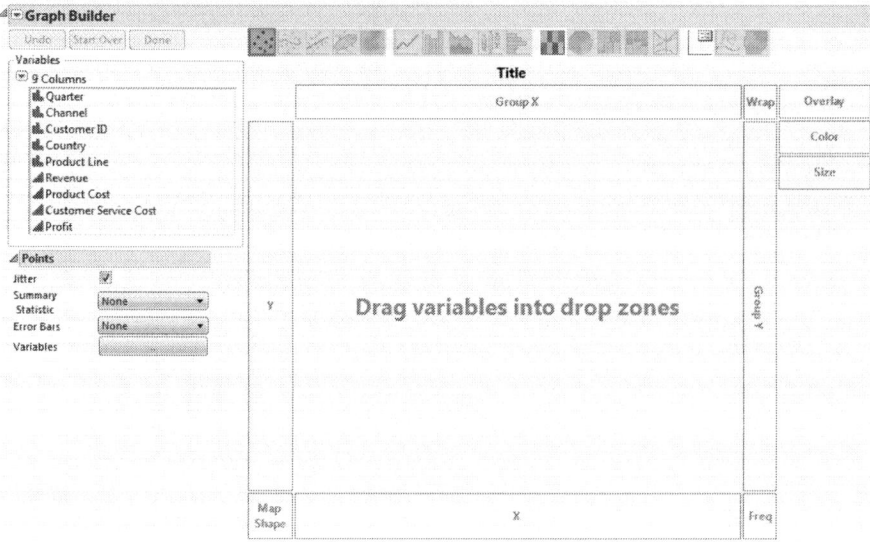

3. Click on Quarter and then drag and drop it onto the X zone to assign **Quarter** as the X variable.

4. Click on **Revenue, Product Cost,** and **Profit,** and drag and drop them onto the Y zone to assign all three variables as Y variables.

 The X and Y zones are now axes.

Note: You can also click on variables and then click a zone to assign them. However, after a zone becomes an axis, drag and drop additional variables onto the axis rather than clicking on the variables and axis.

Figure 3.30 After Adding Y and X Variables

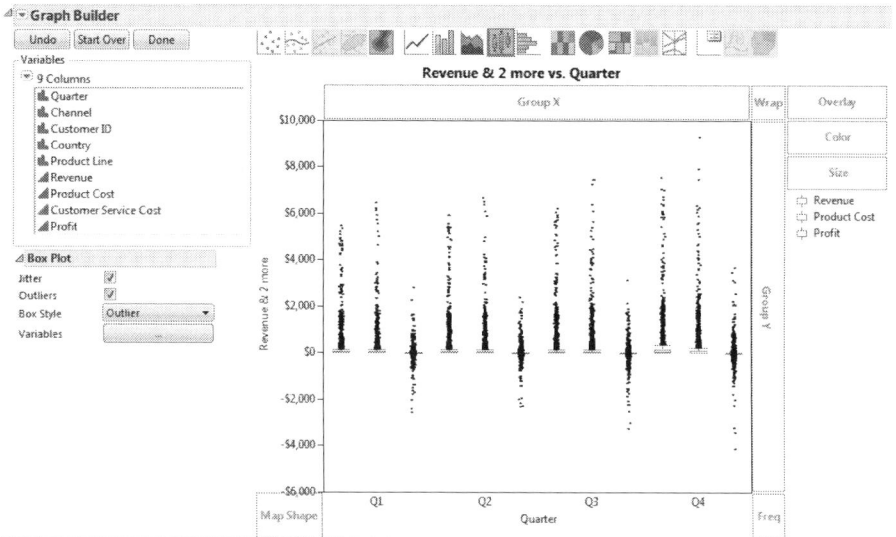

Based on the variables that you are using, Graph Builder shows side-by-side box plots.

5. To change the box plots to a line plot, click on the Line ✓ icon.

Figure 3.31 Line Plot

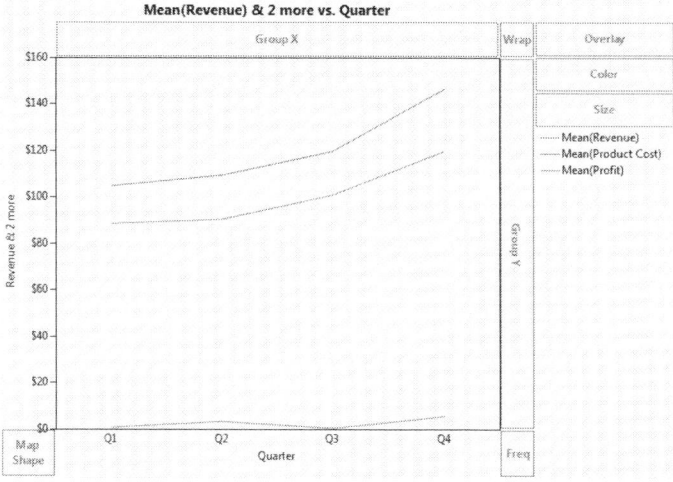

6. To create a separate chart for each product, click on Product Line, and drag and drop it into
 the **Wrap** zone.

 A separate line plot is created for each product.

Figure 3.32 Final Line Plots

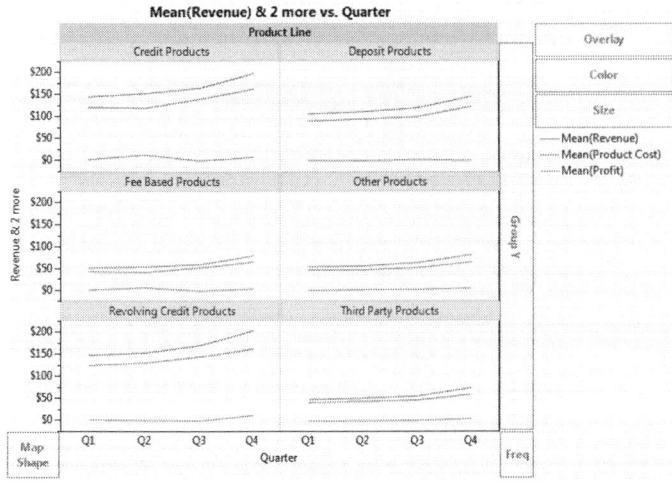

Interpreting the Graph

Figure 3.32 shows revenue, cost, and profit broken down by product line. The business
analyst was interested in seeing the difference in profitability between product lines. The line
plots in Figure 3.32 can provide some answers, as follows:

- Credit products, deposit products, and revolving credit products produce more revenue
 than fee-based products, third-party products, and other products.
- However, the profits of all the product lines are similar.

The data table also includes data on sales channels. The business analyst wants to see how
revenue, product cost, and profit differ between different sales channels.

1. To remove Product Line from the graph, click the title of the graph (Product Line) and drag
 and drop it into any empty area within Graph Builder.
2. To add Channel as the wrap variable, click on Channel and drag and drop it into the **Wrap**
 zone.

Figure 3.33 Line Plots Showing Sales Channels

Figure 3.33 provides this answer: revenue and product cost for ATMs are the highest and are growing the most quickly.

Bubble Plots

A bubble plot is a scatterplot that represents its points as bubbles (or circles). You can change the size and color of the bubbles, and even animate them over time. With the ability to represent up to five dimensions (*x* position, *y* position, size, color, and time), a bubble plot can produce dramatic visualizations and make data exploration easy.

Figure 3.34 Example of a Bubble Plot

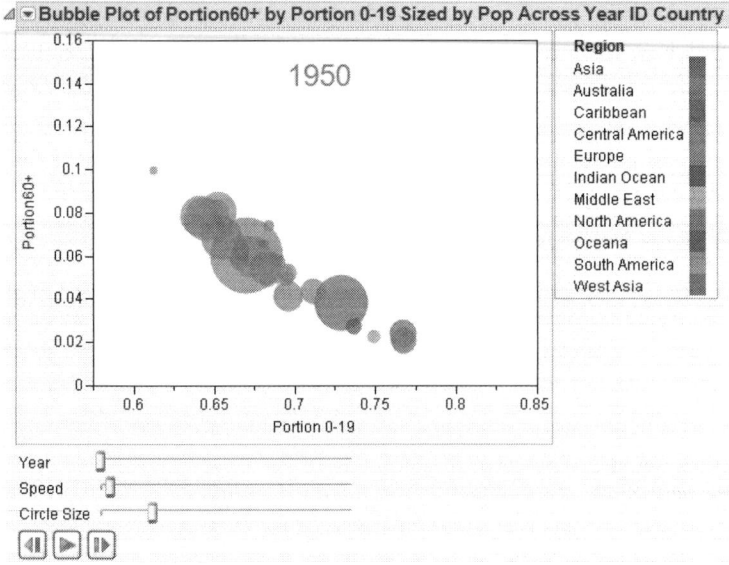

Scenario

This example uses the PopAgeGroup.jmp data table, which contains population statistics for 116 countries or territories between the years 1950 to 2004. Total population numbers are broken out by age group, and not every country has data for every year.

A sociologist wants to explore the following question:

- Is the age of the population of the world changing?

To answer this question, look at the relationship between the oldest (more than 59) and the youngest (younger than 20) portions of the population. Use a bubble plot to determine how this relationship changes over time.

Creating the Bubble Plot

1. Open the PopAgeGroup.jmp data table.
2. Select **Graph** > **Bubble Plot**.
3. Select Portion60+ and click **Y**.

 This corresponds to the *Y* variable on the bubble plot.
4. Select Portion 0-19 and click **X**.

 This corresponds to the *X* variable on the bubble plot.
5. Select Country and click **ID**.

 Each unique level of the ID variable is represented by a bubble on the plot.

6. Select Year and click **Time**.

 This controls the time indexing when the bubble plot is animated.

7. Select Pop and click **Sizes**.

 This controls the size of the bubbles.

8. Select Region and click **Coloring**.

 Each unique level of the Coloring variable is reflected by a bubble of that color. So in this example, all the bubbles for countries located in the same region are the same color.

Figure 3.35 Bubble Plot Window

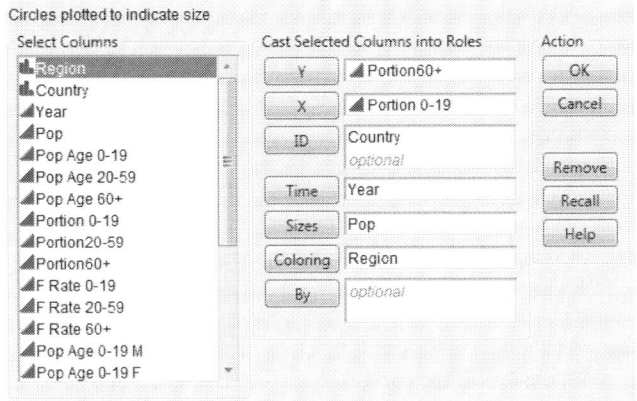

9. Click **OK**.

Figure 3.36 Initial Bubble Plot

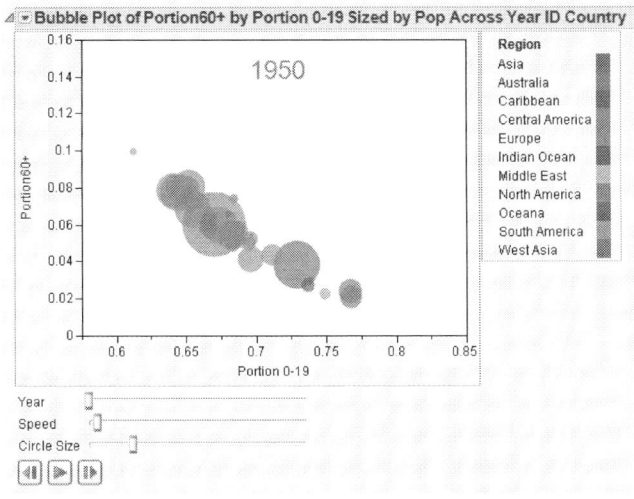

Interpreting the Bubble Plot

Because the time variable (in this case, year) starts in 1950, the initial bubble plot shows the data for 1950. Animate the bubble plot to cycle through all the years by clicking the play/pause button. Each successive bubble plot shows the data for that year. The data for each year determines the following:

- The *X* and *Y* coordinates
- The bubble's sizes
- The bubble's coloring
- Bubble aggregation

Note: For detailed information about how the bubble plot aggregates information across multiple rows, see the *Essential Graphing* book.

The bubble plot for 1950 shows that if a country's proportion of people younger than 20 is high, then the proportion of people more than 59 is low.

Click the play/pause button to animate the bubble plot through the range of years. As time progresses, the Portion 0-19 decreases and the Portion60+ increases.

▶ plays the animation, turns to a pause button after you click it.

❚❚ pauses the animation.

◀❚ manually controls the animation back one unit of time.

❚▶ manually controls the animation forward one unit of time.

Year is used to change the time index manually.

Speed controls the speed of the animation.

Circle Size controls the absolute sizes of the bubbles, while maintaining the relative sizes

The sociologist wanted to know how the age of the world's population is changing. The bubble plot indicates that the population of the world is getting older.

Interacting with the Bubble Plot

Click to select a bubble to see the trend for that bubble over time. For example, in the 1950 plot, the large bubble in the middle is Japan.

To See the Pattern of Population Changes in Japan through the Years

1. Click in the middle of the Japan bubble to select it.
2. From the red triangle menu, select **Trail Bubbles > Selected**.
3. Click the play button.

As the animation progresses through time, the Japan bubble leaves a trail of bubbles that illustrates its history.

Figure 3.37 Japan's History of Population Shifts

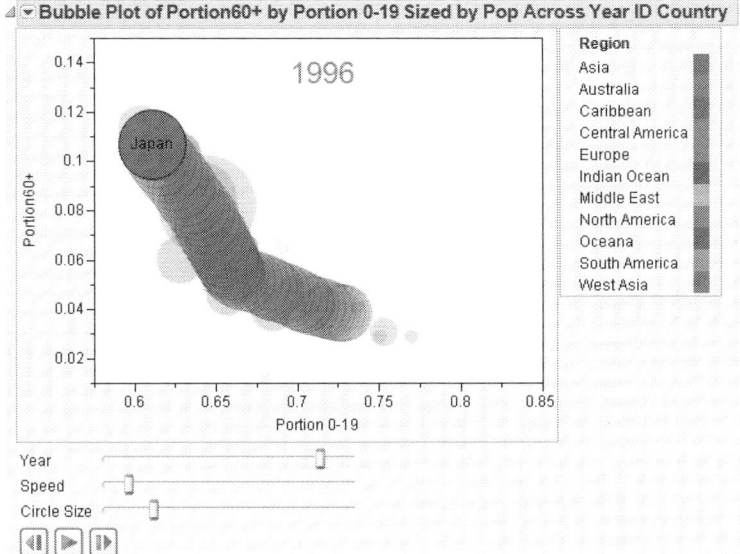

Focusing on the Japan bubble, you can see the following over time:

- The proportion of the population 19 years old or less decreased.
- The proportion of the population 60 years old or more increased.

Chapter **4**

Analyzing Your Data
Distributions, Relationships, and Models

Analyzing your data helps you make informed decisions. Data analysis often involves these actions:

- Examining distributions
- Discovering relationships
- Hypothesis testing
- Building models

Figure 4.1 Analysis Examples

Contents

About This Chapter

Before you analyze your data, review the following information:

- "The Importance of Graphing Your Data" on page 101
- "Understanding Modeling Types" on page 103

The rest of this chapter shows you how to use some basic analytical methods in JMP:

- "Analyzing Distributions" on page 107
- "Analyzing Relationships" on page 112

For a description of advanced modeling and analysis techniques, refer to the following JMP books:

- *Fitting Linear Models*
- *Multivariate Methods*
- *Specialized Models*
- *Consumer Research*
- *Reliability and Survival Methods*
- *Quality and Process Methods*

The Importance of Graphing Your Data

Graphing, or visualizing, your data is important to any data analysis, and should always occur before the use of statistical tests or model building. To illustrate why data visualization should be an early step in your data analysis process, consider the following example:

1. Open the Anscombe.jmp data table (F. J. Anscombe (1973), *American Statistician*, 27, 17-21). This data consists of four pairs of X and Y variables.

2. From the red triangle menu for the The Quartet script in the table panel, select **Run Script**.

 The script creates a simple linear regression on each pair of variables using **Fit Y by X**. The **Show Points** option is turned off, so that none of the data can be seen on the scatterplots. Figure 4.2 shows the model fit and other summary information for each regression.

Figure 4.2 Four Models

Notice that all four models and the RSquare values are nearly identical. The fitted model in each case is essentially $Y = 3 + 0.5X$, and the RSquare value in each case is essentially 0.66. If your data analysis took into account only the above summary information, you would likely conclude that the relationship between X and Y is the same in each case. However, at this point, you have not visualized your data. Your conclusion might be wrong.

To Visualize the Data, Add the Points to All Four Scatterplots

1. Hold down the CTRL key.
2. From the red triangle menu for any one of the Bivariate Fits, select **Show Points**.

Figure 4.3 Scatterplots with Points Added

Plot for
pair 1

Plot for
pair 3

Plot for
pair 2

Plot for
pair 4

The scatterplots show that the relationship between X and Y is not the same for the four pairs, although the lines describing the relationships are the same:

- Plot 1 represents a linear relationship.
- Plot 2 represents a non-linear relationship.
- Plot 3 represents a linear relationship, except for one outlier.
- Plot 4 has all the data at x = 8, except for one point.

This example illustrates that conclusions that are based on statistics alone can be inadequate. A visual exploration of the data should be an early part of any data analysis.

Understanding Modeling Types

In JMP, data can be of different types. JMP refers to this as the modeling type of the data. Table 4.1 describes the three modeling types in JMP.

Table 4.1 Modeling Types

Modeling Type	Description	Examples	Specific Example
Continuous	Numeric data only. Used in operations like sums and means.	Height Temperature Time	The time to complete a test might be 2 hours, or 2.13 hours.
Ordinal	Numeric or character data. Values belong to ordered categories.	Month (1,2,...,12) Letter grade (A, B,...F) Size (small, medium, large)	The month of the year can be 2 (February) or 3 (March), but not 2.13. February comes before March.
Nominal	Numeric or character data. Values belong to categories, but the order is not important.	Gender (M or F) Color Test result (pass or fail)	The gender can be M or F, with no order. Gender categories can also be represented by a number (M=1 and F=2).

Example: Modeling Type Results

Different modeling types produce different results in JMP. To see an example of the differences, follow these steps:

1. Open the Linnerud.jmp sample data table.
2. Select **Analyze** > **Distribution**.
3. Select Age and Weight and click **Y, Columns**.
4. Click **OK**.

Figure 4.4 Distribution Results for Age and Weight

Although Age and Weight are both numeric variables, they are not treated the same. Table 4.2 compares the differences between the results for weight and age.

Table 4.2 Results for weight and age

Variable	Modeling Type	Results
Weight	Continuous	Histogram, Quantiles, and Summary Statistics
Age	Ordinal	Bar chart and Frequencies

Changing the Modeling Type

To treat a variable differently, change the modeling type. For example, in Figure 4.4, the modeling type for Age is ordinal. Remember that for an ordinal variable, JMP calculates frequency counts. Suppose that you wanted to find the average age instead of frequency counts. Change the modeling type to continuous, which shows the mean age.

1. Double-click on the Age column heading. The Column Info window appears.

2. Change the Modeling Type to **Continuous**.

Figure 4.5 Column Info Window

3. Click **OK**.

4. Repeat the steps in the example (see "Example: Modeling Type Results" on page 104) to create the distribution. Figure 4.6 shows the distribution results when Age is ordinal and continuous.

Figure 4.6 Different Modeling Types for age

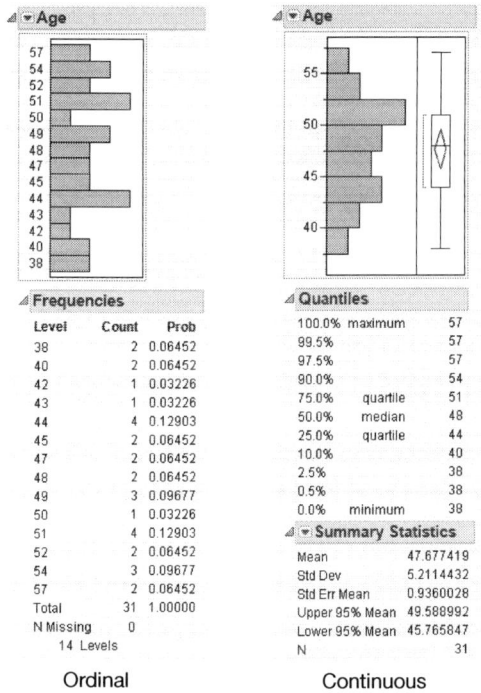

Ordinal Continuous

When age is ordinal, you can see the frequency counts for each age. For example, age 48 appears 2 times. When age is continuous, you can find the mean age, which is nearly 48 (47.677)

Analyzing Distributions

To analyze a single variable, you can examine the distribution of the variable, using the Distribution platform. Report content for each variable varies, depending on whether the variable is categorical (nominal or ordinal) or continuous.

Note: For complete details about the **Distribution** platform, see the *Basic Analysis* book.

Distributions of Continuous Variables

Analyzing a continuous variable might include questions such as the following:

- Does the shape of the data match any known distributions?
- Are there any outliers in the data?
- What is the average of the data?
- Is the average statistically different from a target or historical value?
- How spread out are the data? In other words, what is the standard deviation?
- What are the minimum and maximum values?

You can answer these and other questions with graphs, summary statistics, and simple statistical tests.

Scenario

This example uses the Car Physical Data.jmp data table, which contains information about 116 different car models.

A planning specialist has been asked by a railroad company to determine the possible issues involved in transporting cars by train. Using the data, the planning specialist wants to explore the following questions:

- What is the average car weight?
- How spread out are the cars' weights (standard deviation)?
- What are the minimum and maximum weights of cars?
- Are there any outliers in the data?

Use a histogram of weight to answer these questions.

Creating the Histogram

1. Open the Car Physical Data.jmp sample data table.

2. Select **Analyze** > **Distribution**.

3. Select Weight and click **Y, Columns**.

4. Click **OK**.

5. To rotate the report window, select **Display Options** > **Horizontal Layout** from the red triangle menu next to **Weight**.

Figure 4.7 Distribution of Weight

The report window contains three sections:

- A histogram and a box plot to visualize the data.

- A Quantiles report that shows the percentiles of the distribution.

- A Summary Statistics report that shows the mean, standard deviation, and other statistics.

Interpreting the Distribution Results

Using the results presented in Figure 4.7, the planning specialist can answer the questions.

Table 4.3 Questions Answered Using the Report Window Results[a]

Question	Histogram	Quantiles	Summary Statistics
What is the average car weight?	about 3000 lbs		2958 lbs
How spread out are the weights (standard deviation)?			536 lbs

Table 4.3 Questions Answered Using the Report Window Results[a]

Question	Histogram	Quantiles	Summary Statistics
What are the minimum and maximum weights?	about 1500 lbs and 4500 lbs	1695 lbs and 4285 lbs	
Are there any outliers?	No		

a. A blank cell means not applicable.

The default report window in Figure 4.7 provides a minimal set of graphs and statistics. Additional graphs and statistics are available on the red triangle menu.

Drawing Conclusions

Based on other research, the railroad company has determined that an average weight of 3000 pounds is the most efficient to transport. Now, the planning specialist needs to find out whether the average car weight in the general population of cars that they might transport is 3000 pounds. Use a t-test to draw inferences about the broader population based on this sample of the population.

Testing Conclusions

1. From the red triangle menu for **Weight**, select **Test Mean**.
2. In the window that appears, type 3000 in the Specify Hypothesized Mean box.
3. Click **OK**.

Figure 4.8 Test Mean Results

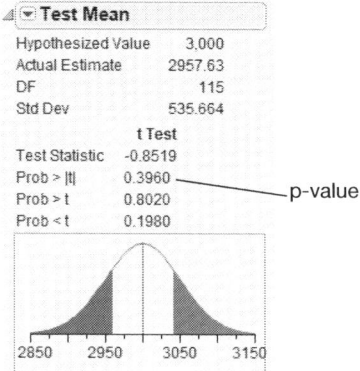

Interpreting the t-Test

The primary result of a t-test is the p-value. In this example, the p-value is 0.396 and the analyst is using a significance level of 0.05. Since 0.396 is greater than 0.05, you cannot conclude that the average weight of car models in the broader population is significantly different from 3000 pounds. Had the p-value been lower than the significance level, the planning specialist would have concluded that the average car weight in the broader population *is* significantly different from 3000 pounds.

Distributions of Categorical Variables

Analyzing a categorical (ordinal or nominal) variable might include questions such as the following:

- How many levels does the variable have?
- How many data points does each level have?
- Is the data uniformly distributed?
- What proportions of the total do each level represent?

Scenario

See the scenario in "Distributions of Continuous Variables" on page 107.

Now that the railroad company has determined that the average weight of the cars is not significantly different from the target weight, there are more questions to address.

The planning specialist wants to answer these questions for the railroad company:

- What are the types of cars?
- What are the countries of origin?

To answer these questions, look at the distribution for Type and Country.

Creating the Distribution

1. Open the Car Physical Data.jmp sample data table.
2. Select **Analyze** > **Distribution**.
3. Select Country and Type and click **Y, Columns**.
4. Click **OK**.

Figure 4.9 Distribution for Country and Type

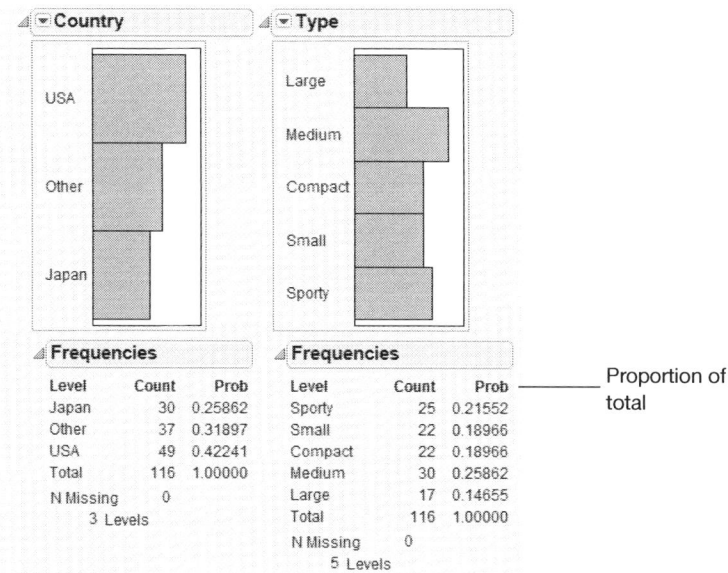

Proportion of
total

Interpreting the Distribution Results

The report window includes a bar chart and a Frequencies report for country and type. The bar chart is a graphical representation of the frequency information provided in the Frequencies report. The Frequencies report contains the following:

- Categories of data. For example, Japan is a category of Country, and Sporty is a category of Type.
- Total counts for each category.
- Proportion of the total each category represents.

For example, there are 22 compact cars, or about 19% of the 116 observations.

Interacting with the Distribution Results

Selecting a bar in one chart also selects the corresponding data in the other chart. For example, select the Japan bar in the Country bar chart to see that a large number of Japanese cars are sporty.

Figure 4.10 Japanese Cars

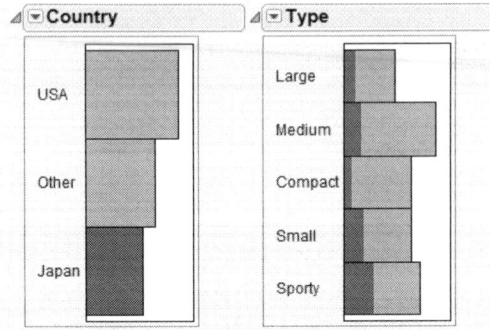

Select the Other category to see that a majority of these cars are small or compact, and almost none are large.

Figure 4.11 Other Cars

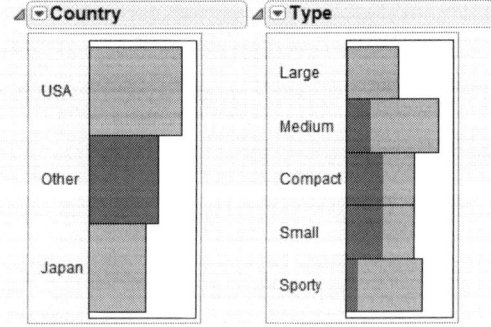

Analyzing Relationships

Scatterplots and other such graphs can help you visualize relationships between variables. Once you have visualized relationships, the next step is to analyze those relationships so that you can describe them numerically. That numerical description of the relationship between variables is called a *model*. Even more importantly, a model also predicts the average value of one variable (*Y*) from the value of another variable (*X*). The *X* variable is also called a predictor. Generally, this model is called a *regression* model.

With JMP, the **Fit Y by X** platform and the **Fit Model** platform creates regression models.

Note: Only the basic platforms and options are covered here. For complete details and explanations of all platform options, see *Basic Analysis, Essential Graphing*, and the books listed in "About This Chapter" on page 101.

Table 4.4 shows the four primary types of relationships.

Table 4.4 Relationship Types

X	Y	Section
Continuous	Continuous	• "Using Regression with One Predictor" on page 113 • "Using Regression with Multiple Predictors" on page 127
Categorical	Continuous	• "Comparing Averages for One Variable" on page 118 • "Comparing Averages for Multiple Variables" on page 123
Categorical	Categorical	"Comparing Proportions" on page 121
Continuous	Categorical	Logistic regression is an advanced topic. See the *Basic Analysis* book.

Using Regression with One Predictor

Scenario

This example uses the Companies.jmp data table, which contains financial data for 32 companies from the pharmaceutical and computer industries.

Intuitively, it makes sense that companies with more employees can generate more sales revenue than companies with fewer employees. A data analyst wants to predict the overall sales revenue for each company based on the number of employees.

To accomplish this task, do the following:

- "Discovering the Relationship" on page 113
- "Fitting the Regression Model" on page 114
- "Predicting Average Sales" on page 116

Discovering the Relationship

First, create a scatterplot to see the relationship between the number of employees and the amount of sales revenue. This scatterplot was created in "Creating the Scatterplot" on page 75 in the "Visualizing Your Data" chapter. After hiding and excluding one outlier (a company with significantly more employees and higher sales), the plot in Figure 4.12 shows the result.

Figure 4.12 Scatterplot of Sales ($M) versus # Employ

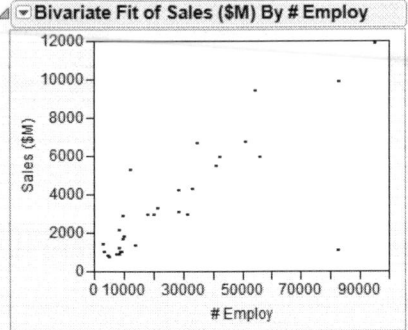

This scatterplot provides a clearer picture of the relationship between sales and the number of employees. As expected, the more employees a company has, the higher sales that it can generate. This visually confirms the data analyst's guess, but it does not predict sales for a given number of employees.

Fitting the Regression Model

To predict the sales revenue from the number of employees, fit a regression model. From the red triangle for **Bivariate Fit**, select **Fit Line**. A regression line is added to the scatterplot and reports are added to the report window.

Figure 4.13 Regression Line

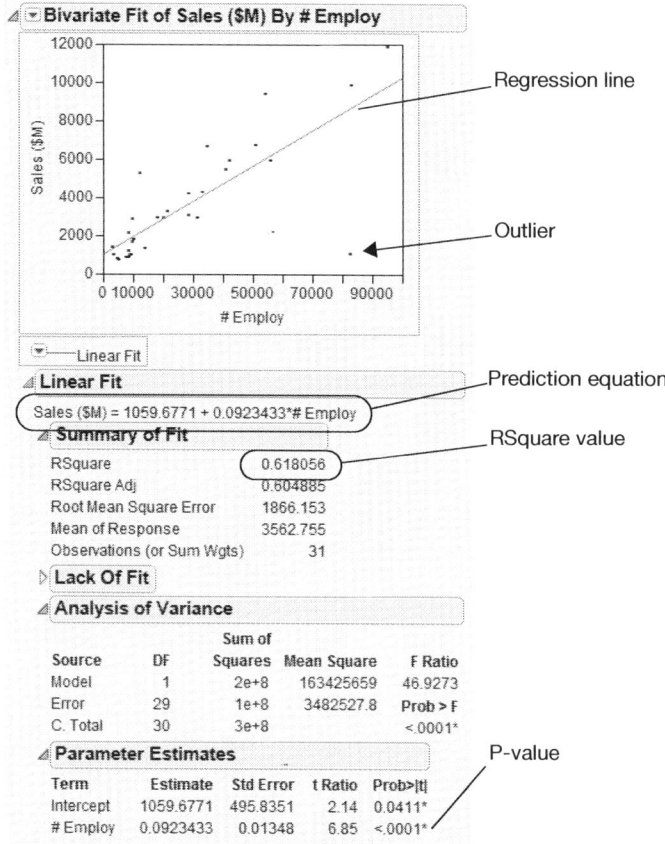

Within the reports, look at the following results:

- the p-value of <.0001

- the RSquare value of 0.618

From these results, the data analyst can conclude the following:

- The p-value is less than the significance level of 0.05. Therefore, including the number of employees in the prediction model significantly improves the ability to predict average sales.

- Since the RSquare value in this example is large, this confirms that a prediction model based on the number of employees can predict sales revenue. The RSquare value shows the strength of a relationship between variables, also called the *correlation*. A correlation of 0 indicates no relationship between the variables, and a correlation of 1 indicates a perfect linear relationship.

Predicting Average Sales

Use the regression model to predict the average sales a company might expect if they have a certain number of employees. The prediction equation for the model is included in the report:

Average sales = 1059.68 + 0.092*employees

For example, in a company with 70,000 employees, the equation is as follows:

$7,499.68 = 1059.68 + 0.092*70,000

In the lower right area of the current scatterplot, there is an outlier that does not follow the general pattern of the other companies. The data analyst wants to know whether the prediction model changes when this outlier is excluded.

Exclude the Outlier

1. Click on the outlier.

2. Select **Rows** > **Exclude/Unexclude**.

3. Fit this model by selecting **Fit Line** from the red triangle menu for **Bivariate Fit**.

The following are added to the report window (see Figure 4.14):

- a new regression line

- a new Linear Fit report, which includes:

 - a new prediction equation

 - a new RSquare value

Figure 4.14 Comparing the Models

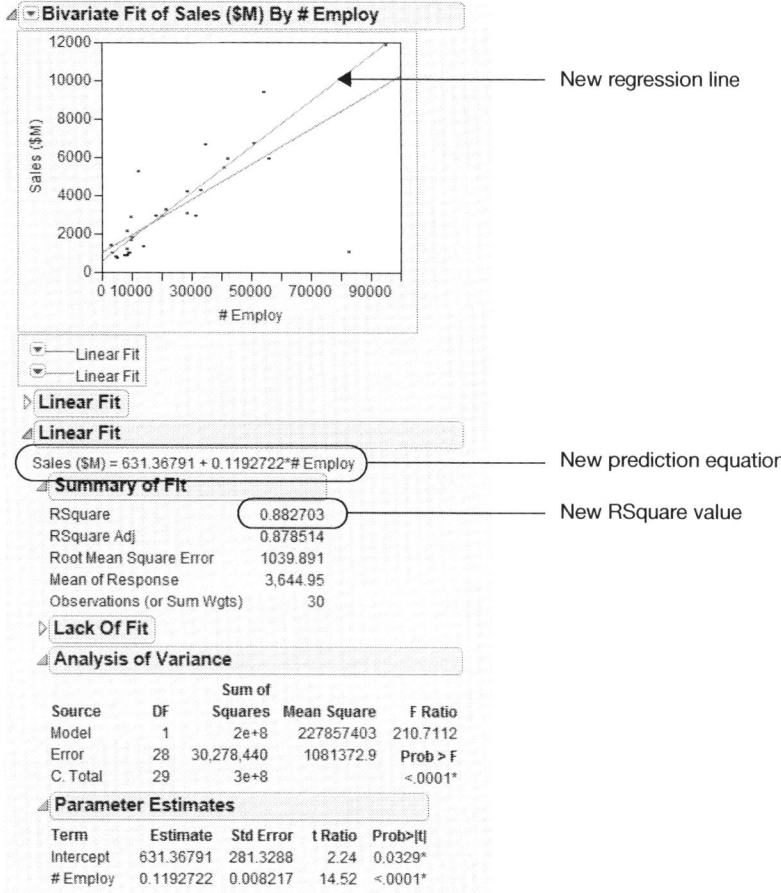

Using the results in Figure 4.14, the data analyst can make the following conclusions:

• The outlier was pulling down the regression line for the larger companies, and pulling the line up for the smaller companies.

• The new model fits the data better, since the new RSquare value (0.88) is closer to 1 than the first RSquare value (0.618).

Using the new prediction equation, the predicted average sales for a company with 70,000 employees can be calculated as follows:

$8961.37 = 631.37 + 0.119*70,000$

The prediction for the first model was $7499.68, so this model predicts a higher sales total by $1461.69.

The second model, after removing the outlier, describes and predicts sales totals based on the number of employees better than the first model. The data analyst now has a good model to use.

Comparing Averages for One Variable

If you have a continuous Y variable, and a categorical X variable, you can compare averages across the levels of the X variable.

Scenario

This example uses the Companies.jmp data table, which contains financial data for 32 companies from the pharmaceutical and computer industries.

A financial analyst wants to explore the following question:

- How do the profits of computer companies compare to the profits of pharmaceutical companies?

To answer this question, fit Profits ($M) by Type.

Discovering the Relationship

1. Open the Companies.jmp sample data table.

2. If you still have the Companies.jmp sample data table open, you might have rows that are excluded or hidden. To return the rows to the default state (all rows included and none hidden), select **Rows > Clear Row States**.

3. Select **Analyze > Fit Y by X**.

4. Select Profits ($M) and click **Y, Response**.

5. Select Type and click **X, Factor**.

6. Click **OK**.

Figure 4.15 Profits by Company Type

There is an outlier in the Computer Type. The outlier is stretching the scale of the plot and making it difficult to compare the profits. Exclude and hide the outlier:

1. Click on the outlier.

2. Select **Rows** > **Exclude/Unexclude**. The data point is no longer included in calculations.

3. Select **Rows** > **Hide/Unhide**. The data point is hidden from all graphs.

4. To recreate the plot without the outlier, select **Script** > **Redo Analysis** from the red triangle menu for Oneway Analysis. You can close the original Scatterplot window.

Figure 4.16 Updated Plot

Removing the outlier gives the financial analyst a clearer picture of the data.

5. To continue analyzing the relationship, select these options from the red triangle menu for Oneway Analysis:

 – **Display Options** > **Mean Lines**. This adds mean lines to the scatterplot.

 – **Means and Std Dev**. This displays a report that provides averages and standard deviations.

Figure 4.17 Mean Lines and Report

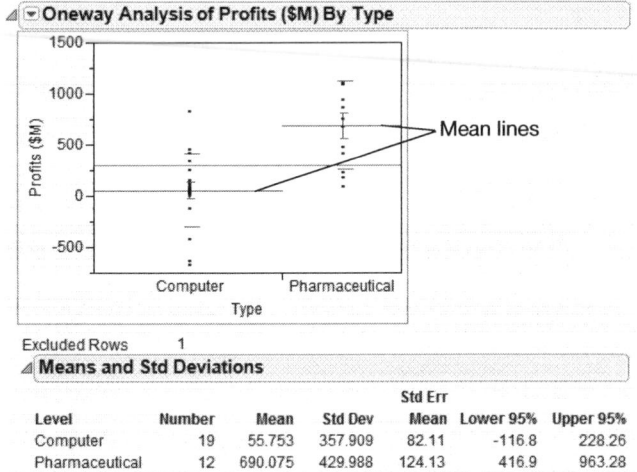

Level	Number	Mean	Std Dev	Std Err Mean	Lower 95%	Upper 95%
Computer	19	55.753	357.909	82.11	-116.8	228.26
Pharmaceutical	12	690.075	429.988	124.13	416.9	963.28

Interpreting the Results

The financial analyst wanted to know how the profits of computer companies compared to the profits of pharmaceutical companies. The updated scatterplot shows that pharmaceutical companies have higher average profits than computer companies. In the report, if you subtract one mean value from the other, the difference in profit is about $635 million. The plot also shows that some of the computer companies have negative profits and all of the pharmaceutical companies have positive profits.

Performing the T-test

The financial analyst has looked at only a sample of companies (the companies in the data table). The financial analyst now wants to examine these questions:

- Does a difference exist in the broader population, or is the difference of $635 million due to chance?

- If there is a difference, what is it?

To answer these questions, perform a two-sample t-test. A t-test lets you use data from a sample to make inferences about the larger population.

To perform the t-test, select **Means/Anova/Pooled t** from the red triangle for Oneway Analysis.

Figure 4.18 t Test Results

Confidence interval limits

p-value

The p-value of 0.0001 is less than the significance level of 0.05, which indicates statistical significance. Therefore, the financial analyst can conclude that the difference in average profits for the sample data is not due to chance alone. This means that in the larger population, the average profits for pharmaceutical companies are different from the average profits for computer companies.

Use the confidence interval limits to determine how much difference exists in the profits of both types of companies. Look at the **Upper CL Dif** and **Lower CL Dif** values in Figure 4.18. Using these values, the financial analyst concludes that the average profit of pharmaceutical companies is between $343 million and $926 million higher than the average profit of computer companies.

Comparing Proportions

If you have categorical X and Y variables, you can compare the proportions of the levels within the Y variable to the levels within the X variable.

Scenario

This example continues to use the Companies.jmp data table. In "Comparing Averages for One Variable" on page 118, a financial analyst determined that pharmaceutical companies have higher profits on average than do computer companies.

The financial analyst wants to know whether the size of a company affects profits more for one type of company than the other? However, before examining this question, the financial analyst needs to know whether the populations of computer and pharmaceutical companies consist of the same proportions of small, medium, and big companies.

Discovering the Relationship

1. Open the Companies.jmp sample data table.

2. If you still have the Companies.jmp data file open from the previous example, you might have rows that are excluded or hidden. To return the rows to the default state (all rows included and none hidden), select **Rows > Clear Row States**.

3. Select **Analyze > Fit Y by X**.

4. Select Size Co and click **Y, Response**.

5. Select Type and click **X, Factor**.

6. Click **OK**.

Figure 4.19 Company Size by Company Type

The Contingency Table contains information that is not applicable for this example. From the red triangle menu for **Contingency Table** deselect **Total %** and **Col %** to remove that information. Figure 4.20 shows the updated table.

Figure 4.20 Updated Contingency Table

		Size Co			
Count	big	medium	small		
Row %					
Computer	4	2	14	20	
	20.00	10.00	70.00		
Pharmaceutical	5	5	2	12	
	41.67	41.67	16.67		
	9	7	16	32	

Interpreting the Results

The statistics in the Contingency Table are graphically represented in the Mosaic Plot. Together, the Mosaic Plot and the Contingency Table compare the percentages of small, medium, and big companies between the two industries. For example, the Mosaic Plot shows that the computer industry has a higher percentage of small companies compared to the pharmaceutical industry. The Contingency Table shows the exact statistics: 70% of computer companies are small, and about 17% of pharmaceutical companies are small.

Interpreting the Test

The financial analyst has looked at only a sample of companies (the companies in the data table). The financial analyst needs to know whether the percentages differ in the broader populations of all computer and pharmaceutical companies.

To answer this question, use the p-value from the Pearson test in the **Tests** report. See Figure 4.19. Since the p-value of 0.011 is less than the significance level of 0.05, the financial analyst concludes the following:

- The differences in the sample data are not due to chance alone.

- The percentages differ in the broader population.

Now the financial analyst knows that the proportions of small, medium, and big companies are different, and can answer the question: Does the size of company affect profits more for one type of company than the other?

Comparing Averages for Multiple Variables

The section "Comparing Averages for One Variable" on page 118, compared averages across the levels of a categorical variable. To compare averages across the levels of two or more variables at once, use the *Analysis of Variance* technique (or ANOVA).

Scenario

The financial analyst can answer the question that we started to work through in the Comparing Proportions section, which is: Does the size of the company have a larger effect on the company's profits, based on type (pharmaceutical or computer)?

To answer this question, compare the company profits by these two variables:

- Type (pharmaceutical or computer)
- Size (small, medium, big)

Discovering the Relationship

To visualize the differences in profit for all of the combinations of type and size, use a graph:

1. Open the Companies.jmp sample data table.
2. Select **Graph > Graph Builder**. The Graph Builder window appears.
3. Click on Profits ($M) and drag and drop it into the **Y** zone.
4. Click on Size Co and drag and drop it into the **X** zone.
5. Click on Type and drag and drop it into the **Group X** zone.

Figure 4.21 Graph of Company Profits

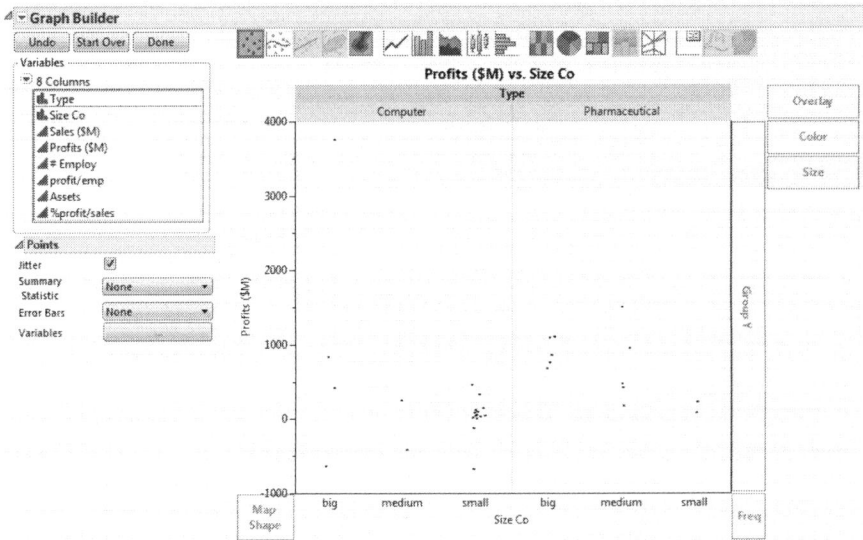

The graph shows that one big computer company has very large profits. That outlier is stretching the scale of the graph, making it difficult to compare the other data points.

6. Right-click on the outlier to select it, and then select **Row Exclude**. The point is removed, and the scale of the graph automatically updates.

7. Click on the Bar icon. Comparing mean profits is easier with bar charts than with points.

Figure 4.22 Graph with Outlier Removed

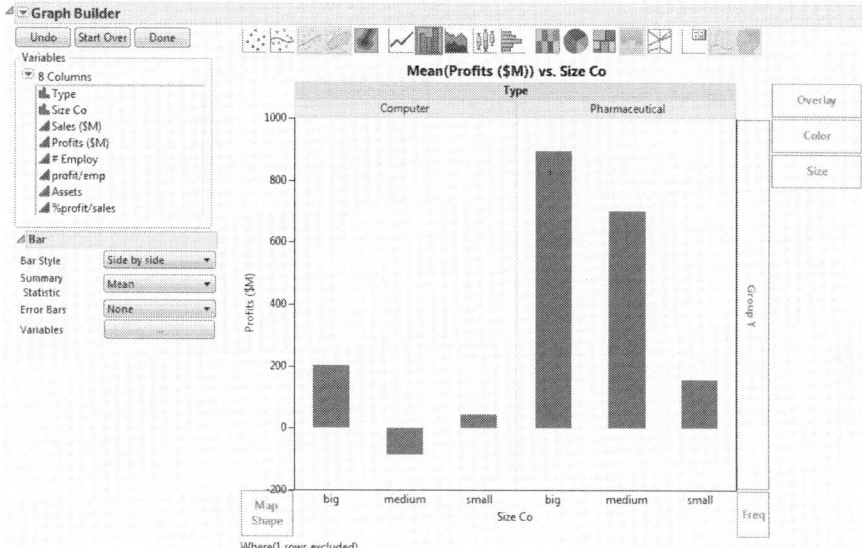

The updated graph shows that pharmaceutical companies have higher average profits. The graph also shows that profits differ between company sizes for only the pharmaceutical companies. When the effect of one variable (company size) changes for different levels of another variable (company type), this is called an *interaction*.

Quantifying the Relationship

Since this data is only a sample, the financial analyst needs to determine the following:

• if the differences are limited to this sample and due to chance

 or

• if the same patterns exist in the broader population

1. Return to the Companies.jmp sample data table that has the data point excluded. See "Discovering the Relationship" on page 124.

2. Select **Analyze > Fit Model**.

3. Select Profits ($M) and click **Y**.

4. Select both Type and Size Co.

5. Click the **Macros** button and select **Full Factorial**.

6. From the Emphasis menu, select **Effect Screening**.

7. Select the **Keep dialog open** option.

Figure 4.23 Completed Fit Model Window

8. Click **Run**. The report window shows the model results.

To decide whether the differences in profits are real, or due to chance, examine the **Effect Tests** report.

Note: For complete details about all of the **Fit Model** results, see the *Fitting Linear Models* book.

Using Effect Tests

The Effect Tests report (see Figure 4.24) shows the results of the statistical tests. There is a test for each of the effects included in the model on the Fit Model window: Type, Size Co, and Type*Size Co.

Figure 4.24 Effect Tests Report

Source	Nparm	DF	Sum of Squares	F Ratio	Prob > F
Type	1	1	1401847.4	10.1368	0.0039*
Size Co	2	2	724616.2	2.6198	0.0927
Type*Size Co	2	2	448061.5	1.6200	0.2180

First, look at the test for the interaction in the model: the Type*Size Co effect. Figure 4.22 showed that the pharmaceutical companies appeared to have different profits between company sizes. However, the effect test indicates that there is no interaction between type and

Chapter 4
Discovering JMP

Analyzing Your Data 127
Analyzing Relationships

size as it relates to profit. The p-value of 0.218 is large (greater than the significance level of 0.05). Therefore, remove that effect from the model, and re-run the model.

1. Return to the Fit Model window.

2. In the Construct Model Effects box, select the **Type*Size Co** effect and click **Remove**.

3. Click **Run**.

Figure 4.25 Updated Effect Tests Report

Source	Nparm	DF	Sum of Squares	F Ratio	Prob > F
Type	1	1	1,356.298	9.3768	0.0049*
Size Co	2	2	434,161.3	1.5008	0.2410

The p-value for the Size Co effect is large, indicating that there are no differences based on size in the broader population. The p-value for the Type effect is small, indicating that the differences that you saw in the data between computer and pharmaceutical companies is not due to chance.

Conclusions

The financial analyst wanted to know whether the size of the company has a larger effect on the company's profits, based on type (pharmaceutical or computer).The financial analyst can now answer the question as follows:

- There is a real difference in profits between computer and pharmaceutical companies in the broader population.

- There is no correlation between the company's size and type and its profits.

Using Regression with Multiple Predictors

The section "Using Regression with One Predictor" on page 113 showed you how to build simple regression models consisting of one predictor variable and one response variable. *Multiple regression* predicts the average response variable using two or more predictor variables.

Scenario

This example uses the Candy Bars.jmp data table, which contains nutrition information for candy bars.

A dietician wants to predict calories using the following information:

- Total fat

- Carbohydrates

- Protein

Use *multiple regression* to predict the average response variable using these three predictor variables.

Discovering the Relationship

To visualize the relationship between calories and total fat, carbohydrates, and protein, create a scatterplot matrix:

1. Open the Candy Bars.jmp sample data table.
2. Select **Graph** > **Scatterplot Matrix**.
3. Select Calories and click **Y, Columns**.
4. Select Total fat g, Carbohydrate g, and Protein g, and click **X**.
5. Click **OK**.

Figure 4.26 Scatterplot Matrix Results

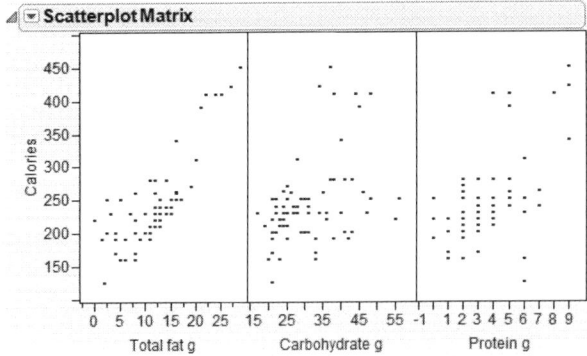

The scatterplot matrix shows that there is a positive correlation between calories and all three variables. The correlation between calories and total fat is the strongest. Now that the dietician knows that there is a relationship, the dietician can build a multiple regression model to predict average calories.

Building the Multiple Regression Model

Continue to use the Candy Bars.jmp sample data table.

1. Select **Analyze** > **Fit Model**.
2. Select Calories and click **Y**.
3. Select Total Fat g, Carbohydrate g, and Protein g and click **Add**.
4. Next to Emphasis, select **Effect Screening**.

Figure 4.27 Fit Model Window

5. Click **Run**.

The report window shows the model results. To interpret the model results, focus on these areas:

- "Using the Actual by Predicted Plot" on page 129
- "Interpreting the Parameter Estimates" on page 130
- "Using the Prediction Profiler" on page 131

Note: For complete details about all of the model results, see the *Fitting Linear Models* book.

Using the Actual by Predicted Plot

The Actual by Predicted Plot shows the actual calories versus the predicted calories. As the predicted values come closer to the actual values, the points on the scatterplot fall closer around the red line. See Figure 4.28. Because the points are all very close to the line, you can see that the model predicts calories based on the chosen factors well.

Figure 4.28 Actual by Predicted Plot

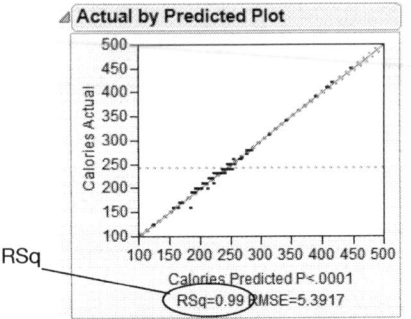

Another measure of model accuracy is the RSq value (which appears below the plot in Figure 4.28). The RSq value measures the percentage of variability in calories, as explained by the model. A value closer to 1 means a model is predicting well. In this example, the RSq value is 0.99.

Interpreting the Parameter Estimates

The Parameter Estimates report shows the following information:

- The model coefficients

- P-values for each parameter

Figure 4.29 Parameter Estimates Report

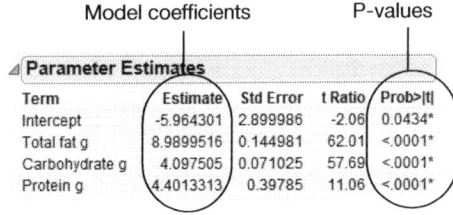

In this example, the p-values are all very small (<.0001). This indicates that all three effects (fat, carbohydrate, and protein) contribute significantly when predicting calories.

You can use the model coefficients to predict the value of calories for particular values of fat, carbohydrate, and protein. For example, suppose that you want to predict the average calories for any candy bar that has these characteristics:

- Fat = 11 g

- Carbohydrate = 43 g

- Protein = 2 g

Using these values, you can calculate the predicted average calories as follows:

$$277.92 = -5.9643 + 8.99*11 + 4.0975*43 + 4.4013*2$$

The characteristics in this example are the same as the Milky Way candy bar (on row 59 of the data table). The actual calories for the Milky Way are 280, showing that the model predicts well.

Using the Prediction Profiler

Use the Prediction Profiler to see how changes in the factors affect the predicted values. The profile lines show the magnitude of change in calories as the factor changes. The line for Total fat g is the steepest, meaning that changes in total fat have the largest effect on calories.

Figure 4.30 Prediction Profiler

Click and drag the vertical line for each factor to see how the predicted value changes. You can also click the current factor values and change them. For example, click on the factor values and type the values for the Milky Way candy bar (row 59).

Figure 4.31 Factor Values for the Milky Way

Note: For complete details about the Prediction Profiler, see the Profilers book.

Conclusion

The dietician now has a good model to predict calories of a candy bar based on its total fat, carbohydrates, and protein.

Chapter **5**

Saving and Sharing Your Work

Saving and Recreating Results

Once you have generated results from your data, JMP provides you with multiple ways to share your work with others. Here are some of the ways that you can share your work:

- Saving platform results as journals or projects
- Saving results, data tables, and other files in projects
- Saving scripts to reproduce results in data tables
- Creating Adobe Flash (.SWF) versions of platform results

Figure 5.1 Example of an Adobe Flash (.SWF) Bubble Plot

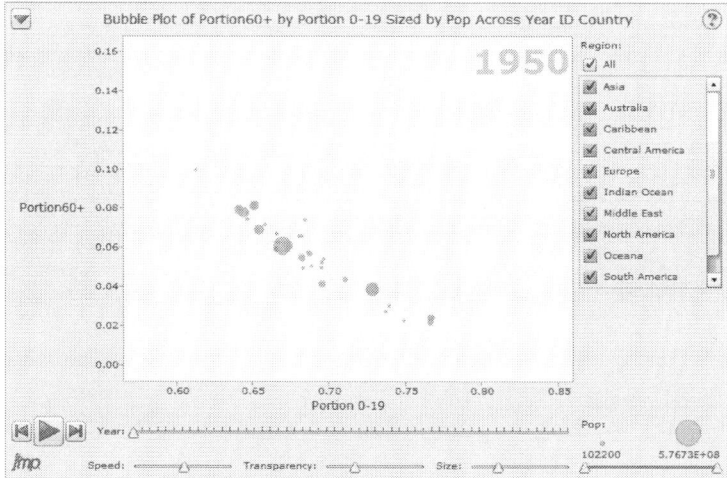

Contents

Saving Platform Results in Journals

Save platform reports for future viewing by creating a journal of the report window. The journal is a copy of the report window. You can edit or append additional reports to an existing journal. The journal is not connected to the data table. A journal is an easy way to save the results from several report windows in a single report window that you can share with others.

Example: Creating a Journal

Create and Modify a Report and Save It to a Journal

1. Open the Companies.jmp sample data table.
2. Select **Analyze** > **Distribution**.
3. Select both Type and Size Co and click **Y, Columns**.
4. Click **OK**.
5. From the red triangle menu for **Type**, select **Histogram Options** > **Show Counts**.
6. From the red triangle menu for **Size Co**, select **Mosaic Plot**.
7. Select **Edit** > **Journal** to journal these results. The results are duplicated in a journal window.

Figure 5.2 Journal of Distribution Results

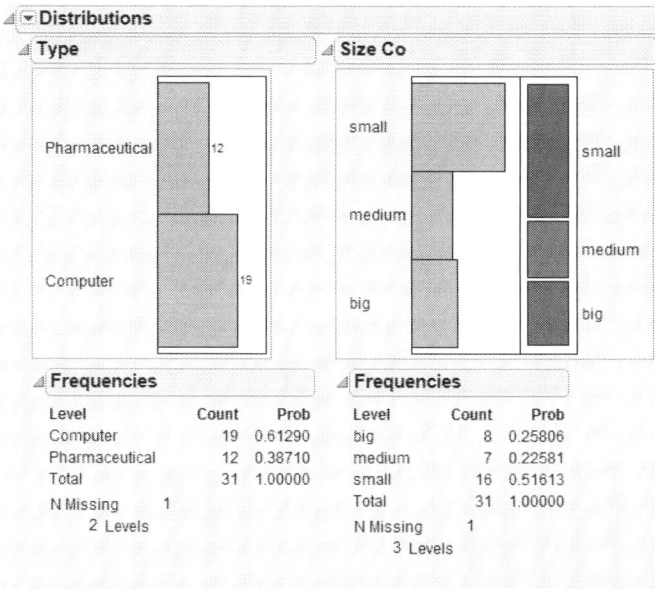

The results in the journal are not connected to the data table. In the Type bar chart, if you click on the Computer bar, no rows are selected in the data table.

Since the journal is a copy of your results, most of the red triangle menus do not exist. A journal does have a red triangle menu for each new report that you add to the journal. This menu has two options:

Rerun in new window If you have the original data table that was used to create the original report, this option runs the analysis again. The result is a new report window.

Edit Script This option opens a script window that contains a JSL script to recreate the analysis. JSL is a more advanced topic that is covered in the *Scripting Guide*.

Adding Additional Analyses

If you perform another analysis, you can add the results of the analysis to the existing journal.

1. With a journal open, select **Analyze** > **Distribution**.

2. Select profit/emp and click **Y, Columns**.

3. Click **OK**.

4. Select **Edit** > **Journal**. The results are appended to the bottom of the journal.

Creating Projects

Save multiple JMP file types (such as data tables, reports, journals, and scripts) in a single file by creating a project. The project file contains all the information needed to re-open all of the included files.

Example: Creating a Project

Create a Report

1. Open the Companies.jmp sample data table.

2. Select **Analyze** > **Distribution**.

3. Select Profits ($M) and profit/emp, and click Y, Columns.

4. Click **OK**.

Create a Project and Add the Data Table and the Report to It

1. To start a new project, select **File** > **New** > **Project**. A new project appears in the Home Window.

Figure 5.3 Initial Project Window on Windows

2. Right-click the project (Untitled) and select **Rename**, and then enter a new name (Finance).

3. To add the Distribution results to the project, right-click on the project name and select **Add Window.**

4. In the Add Windows to Project window, select the Distribution results.

5. Click **OK.** The Distribution results are added to the project.

6. To add the Companies.jmp data table to the project, repeat step 3 and select the Companies data table from the window.

7. Click **OK.** The data table is added to the project.

Figure 5.4 The Final Project

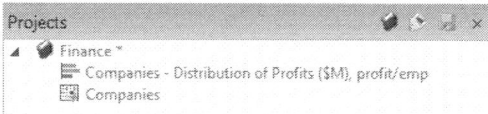

You can double-click on the links in the project to open the data table and recreate the Distribution results.

Using Scripts

Everything in JMP is scriptable, meaning that any action that you perform can be saved as a JMP Scripting Language (JSL) script. You can use a script to reproduce your actions or results at any time.

Example: Saving and Running a Script

Create a Report

1. Open the Companies.jmp sample data table.

2. Select **Analyze** > **Distribution**.

3. Select Type and profit/emp and click **Y, Columns.**

4. Click **OK.**

5. From the red triangle menu for Type, select these options:

 – **Histogram Options** > **Show Counts**

 – **Confidence Interval** > **0.95**

6. From the red triangle menu for profit/emp, select these options:

 – **Outlier Box Plot**, to remove the outlier box plot

 – **CDF Plot**

7. From the red triangle menu for Distributions, select **Stack**.

Save the Script to the Data Table and Run It

1. To save this analysis, select **Script** > **Save Script to Data Table** from the red triangle menu for Distributions. The Distribution script appears in the Table panel.

Figure 5.5 Distribution Script

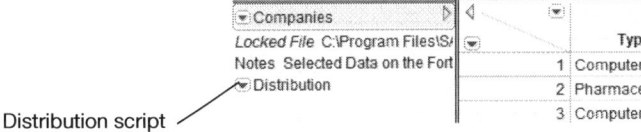

Distribution script

2. Close the Distribution report window.

3. To recreate the analysis, select **Run Script** from the red triangle menu for the Distribution script.

Figure 5.6 Running the Distribution Script

About Scripts and JSL

The script that you saved in this section contains JMP Scripting Language (JSL) commands. JSL is a more advanced topic that is covered in the *Scripting Guide*.

Creating Adobe Flash Versions of the Profiler, Bubble Plot, or Distribution Platforms

To share the interactive results of a profiler, bubble plot, or distribution outside of JMP, you can export a .SWF file and use the Adobe Flash Player to view it. You can import the .SWF file into presentations and applications. You can also save the results as an HTML page with the .SWF output embedded.

Example: Saving an Adobe Flash Version of a Bubble Plot

Create a Bubble Plot in JMP

1. Open the PopAgeGroup.jmp sample data table.

2. Select **Graph > Bubble Plot**.

3. Select Portion60+ and click **Y**.

4. Select Portion 0-19 and click **X**.

5. Select Country and click **ID**.

6. Select Year and click **Time**.

7. Select Pop and click **Sizes**.

8. Select Region and click **Coloring**.

9. Click **OK**.

Figure 5.7 Initial Bubble Plot

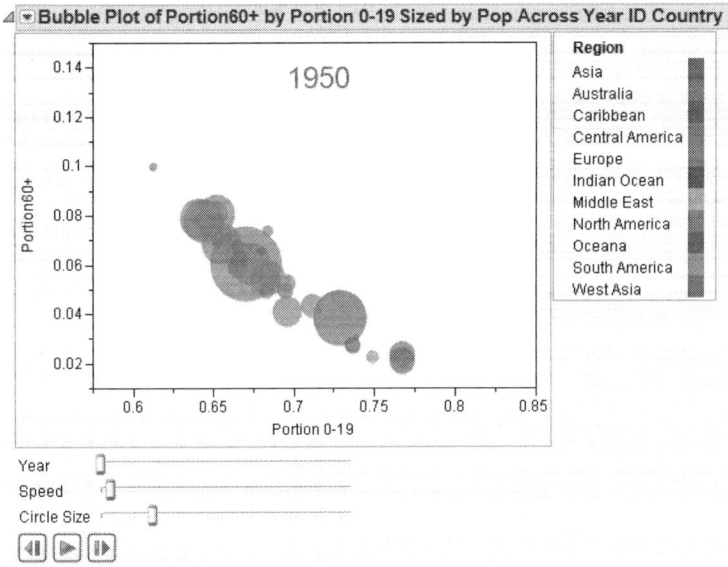

Save the Bubble Plot as a .SWF File

1. Select **Save for Adobe Flash Platform (.SWF)** from the red triangle menu.

2. In the Save As SWF window, select the location to which you want to save the file.

3. Click **Save**. The Adobe Flash version of the Bubble Plot is saved as HTML and appears in a Web browser.

Chapter 5
Discovering JMP

Saving and Sharing Your Work 141
Creating Adobe Flash Versions of the Profiler, Bubble Plot, or Distribution Platforms

Figure 5.8 Adobe Flash Version of Bubble Plot

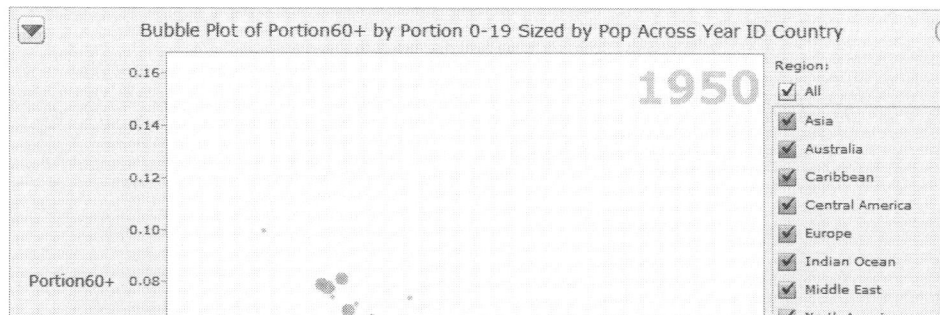

Additional Information

The JMP Web site has additional information about the following:

- more details about Adobe Flash versions of the Profiler, Bubble Plot, and Distribution platforms

- instructions on importing the Adobe Flash versions into Microsoft PowerPoint

This information is on the JMP Web site here:

www.jmp.com/support/swfhelp/en/

Chapter **6**

Special Features
Automatic Updating and Integrating with SAS

Using some of the special features in JMP, you can do the following:

- Update analyses or graphs automatically
- Customize platform results
- Integrate with SAS to use advanced analytical features

Figure 6.1 Examples of Special Features

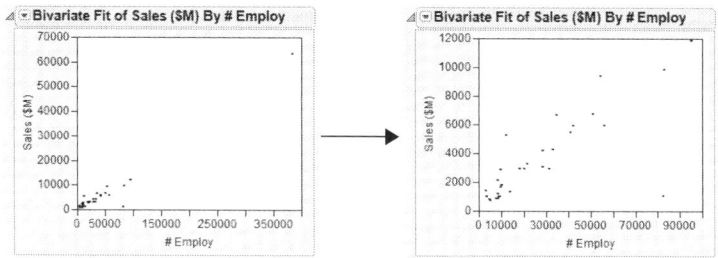

```
DATA Candy_Bars; INPUT  Calories Total_fat_g Carbohydrate_g Protein_g; Lines;
310   20   28   6
230   12   27   4
220   12   24   3
170   8    21   3
200   2.5  43   1
260   16   26   5
410   25   38   8
240   14   22   7
210   13   25   2
230   3    47   3
;
RUN;

PROC GLM  DATA=Candy_Bars ALPHA=0.05;
MODEL  Calories =  Total_fat_g Carbohydrate_g Protein_g;
RUN;
```

Contents

Automatically Updating Analyses and Graphs

When you make a change to a data table, you can use the Automatic Recalc feature to automatically update analyses and graphs that are associated with the data table. For example, if you exclude, include, or delete values in the data table, that change is instantly reflected in the associated analyses or graphs. Note the following information:

- Some platforms do not support Automatic Recalc. For more information, see the *Using JMP* book.

- For the supported platforms in the **Analyze** menu, Automatic Recalc is turned off by default. However, for the supported platforms in the **Quality and Process** menu, Automatic Recalc is turned on by default, except for the Variability/Attribute Gauge Chart, Capability, and Control Chart.

- For the supported platforms in the **Graph** menu, Automatic Recalc is turned on by default.

Example: Using Automatic Recalc

This example uses the Companies.jmp sample data table, which contains financial data for 32 companies from the pharmaceutical and computer industries.

1. Open the Companies.jmp sample data table.
2. Select **Analyze** > **Fit Y by X**.
3. Select Sales ($M) and click **Y, Response**.
4. Select # Employ and click **X, Factor**.
5. Click **OK**.

Figure 6.2 Initial Scatterplot

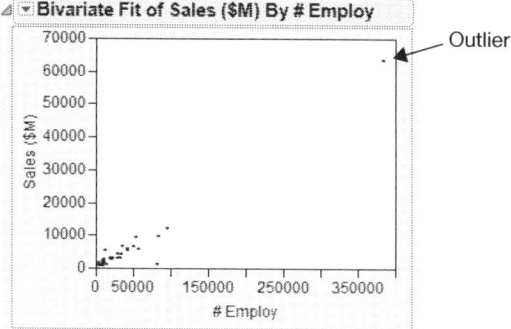

The initial scatterplot shows that one company has significantly more employees and sales than the other companies. You decide that this company is an outlier, and you want to exclude

that point. Before you exclude the point, turn on Automatic Recalc so that your scatterplot is updated automatically when you make the change.

6. Turn on Automatic Recalc by selecting **Script** > **Automatic Recalc** from the red triangle menu.

7. Click on the outlier to select it.

8. Select **Rows** > **Exclude/Unexclude**. The point is excluded from the analysis and the scatterplot is automatically updated.

Figure 6.3 Updated Scatterplot

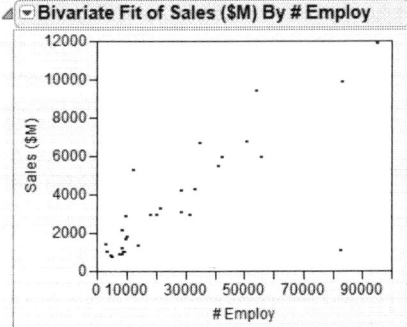

If you fit a regression line to the data, the point in the lower right corner is an outlier, and influences the slope of the line. If you then exclude the outlier with Automatic Recalc turned on, you can see the slope of the line change.

9. Fit a regression line by selecting **Fit Line** from the red triangle menu. Figure 6.4 shows the regression line and analysis results added to the report window.

Figure 6.4 Regression Line and Analysis Results

10. Click on the outlier to select it.

11. Select **Rows > Exclude/Unexclude**. The regression line and analysis results are automatically updated, reflecting the exclusion of the point.

Tip: When you exclude a point, the analyses are recalculated without the data point, but the data point is not hidden in the scatterplot. To also hide the point in the scatterplot, select the point, and then select **Rows > Hide and Exclude**.

Figure 6.5 Updated Regression Line and Analysis Results

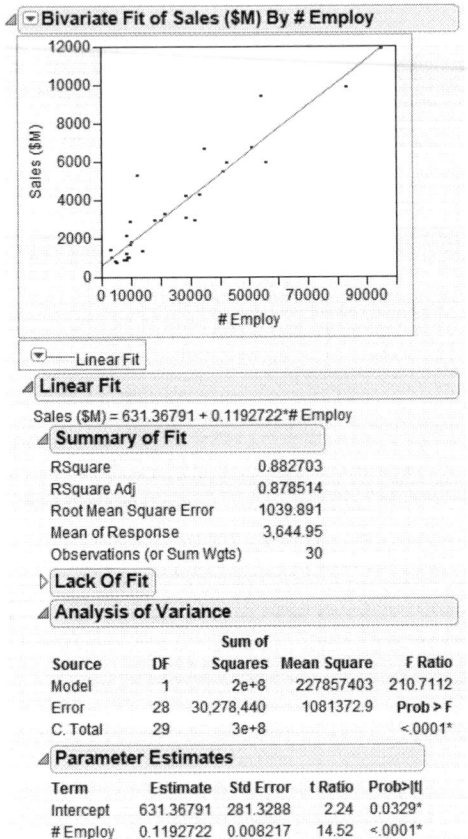

Sales ($M) = 631.36791 + 0.1192722*# Employ

Summary of Fit

RSquare	0.882703
RSquare Adj	0.878514
Root Mean Square Error	1039.891
Mean of Response	3,644.95
Observations (or Sum Wgts)	30

▷ **Lack Of Fit**

Analysis of Variance

Source	DF	Sum of Squares	Mean Square	F Ratio
Model	1	2e+8	227857403	210.7112
Error	28	30,278,440	1081372.9	Prob > F
C. Total	29	3e+8		<.0001*

Parameter Estimates

| Term | Estimate | Std Error | t Ratio | Prob>|t| |
|---|---|---|---|---|
| Intercept | 631.36791 | 281.3288 | 2.24 | 0.0329* |
| # Employ | 0.1192722 | 0.008217 | 14.52 | <.0001* |

Changing Preferences

You can change preferences in JMP using the Preferences window. To open the Preferences window, select **File** > **Preferences**.

Figure 6.6 Preferences Window

On the left side of the Preferences window is a list of Preference groups. On the right side of the window are all of the preferences that you can change for the selected category.

Example: Changing Preferences

Every platform report window has options that you can turn on or off. However, your changes to these options are not remembered the next time you use the platform. If you want JMP to remember your changes every time you use the platform, change those options in the Preferences window.

This example shows how to set the Distribution platform so that an Outlier Box Plot is not added to the initial report.

Create a Distribution Using the Default Preference Setting

1. Open the Companies.jmp sample data table.
2. Select **Analyze** > **Distribution**.
3. Select Profits ($M) and click **Y, Columns**.
4. Click **OK**.

Figure 6.7 Distribution Report Window

The histogram is vertical, and the graphs includes an outlier box plot. To change the histogram to horizontal and remove the outlier box, select the appropriate options from the red triangle menu for Profits ($M). However, if you want those preferences to be in effect every time you use the platform, then change them in the Preferences window.

Change the Preference for the Outlier Box Plot and Run Distribution Again

1. Select **File** > **Preferences**.
2. Select **Platforms** from the preference group.
3. Select **Distribution** from the Platforms list.
4. Select the **Horizontal Layout** option to turn it on.
5. Deselect the **Outlier Box Plot** option to turn it off.

Figure 6.8 Distribution Preferences

6. Click **OK**.

7. Repeat the Distribution analysis. See "Create a Distribution Using the Default Preference Setting" on page 149.

The histogram is now horizontal and the outlier box plot does not appear. These preferences remain the same until you change them.

For details about all of the preferences, see *Using JMP*.

Integrating JMP and SAS

Note: You must have access to SAS, either on your local machine or on a server, to use SAS through JMP.

Using JMP, you can interact with SAS as follows:

* Write or create SAS code in JMP.

* Submit SAS code and view the results in JMP.

* Connect to a SAS Metadata Server or a SAS Server on a remote machine.

* Connect to SAS on your local machine.

* Open and browse SAS data sets.

* Retrieve and view data sets generated by SAS.

For complete details about integrating JMP and SAS, see the *Using JMP*.

Example: Creating SAS Code

This example uses the Candy Bars.jmp sample data table, which contains nutrition data for candy bars.

1. Open the Candy Bars.jmp sample data table.
2. Select **Analyze** > **Fit Model**.
3. Select Calories and click **Y**.
4. Select Total fat g, Carbohydrate g, and Protein g, and click **Add**.
5. From the red triangle menu for Model Specification, select **Create SAS Job**.

Figure 6.9 shows the SAS code. (Not all of the data is shown.)

Figure 6.9 SAS Code

```
DATA Candy_Bars; INPUT  Calories Total_fat_g Carbohydrate_g Protein_g; Lines;
310   20   28   6
230   12   27   4
220   12   24   3
170   8   21   3
200   2.5   43   1
260   16   26   5
410   25   38   8
240   14   22   7
210   13   25   2
230   3   47   3
;
RUN;

PROC GLM  DATA=Candy_Bars ALPHA=0.05;
MODEL  Calories =  Total_fat_g Carbohydrate_g Protein_g;
RUN;
```

Example: Submitting SAS Code

1. Open the Candy Bars.jmp sample data table.
2. Select **Analyze** > **Fit Model**.
3. Select Calories and click **Y**.
4. Select Total fat g, Carbohydrate g, and Protein g, and click **Add**.
5. From the red triangle menu for Model Specification, select **Submit to SAS**.
6. In the **Connect to SAS Server** window (see Figure 6.10), choose a method to connect to SAS (if you are not already connected). For this example, select **Connect to SAS on this machine**.

Figure 6.10 Connect to SAS Server

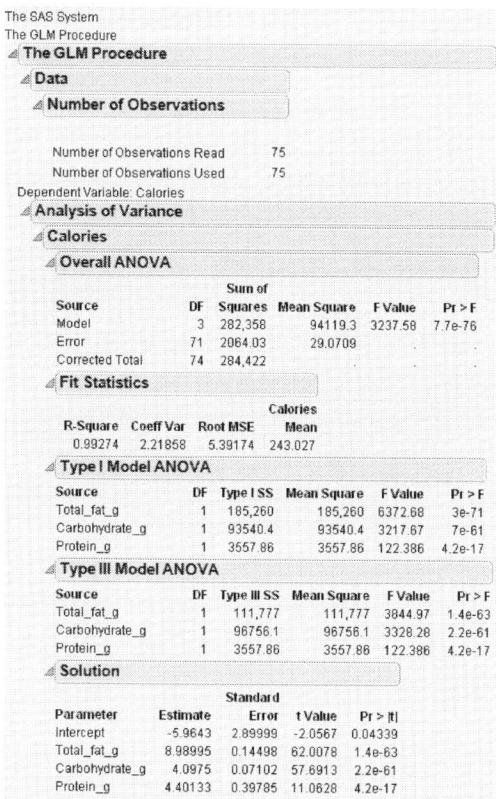

7. Click **OK**.

JMP connects to SAS. SAS runs the model and sends the results back to JMP. The results can appear as SAS output, HTML, RTF, PDF, or JMP report format (you can choose the format using JMP Preferences). Figure 6.11 shows the results formatted as a JMP report. For details, see *Using JMP*.

Figure 6.11 SAS Results Formatted as a JMP Report

The SAS System
The GLM Procedure

◢ The GLM Procedure

◢ Data

◢ Number of Observations

Number of Observations Read	75
Number of Observations Used	75

Dependent Variable: Calories

◢ Analysis of Variance

◢ Calories

◢ Overall ANOVA

Source	DF	Sum of Squares	Mean Square	F Value	Pr > F
Model	3	282,358	94119.3	3237.58	7.7e-76
Error	71	2064.03	29.0709		
Corrected Total	74	284,422			

◢ Fit Statistics

R-Square	Coeff Var	Root MSE	Calories Mean
0.99274	2.21858	5.39174	243.027

◢ Type I Model ANOVA

Source	DF	Type I SS	Mean Square	F Value	Pr > F
Total_fat_g	1	185,260	185,260	6372.68	3e-71
Carbohydrate_g	1	93540.4	93540.4	3217.67	7e-61
Protein_g	1	3557.86	3557.86	122.386	4.2e-17

◢ Type III Model ANOVA

Source	DF	Type III SS	Mean Square	F Value	Pr > F
Total_fat_g	1	111,777	111,777	3844.97	1.4e-63
Carbohydrate_g	1	96756.1	96756.1	3328.28	2.2e-61
Protein_g	1	3557.86	3557.86	122.386	4.2e-17

◢ Solution

| Parameter | Estimate | Standard Error | t Value | Pr > |t| |
|---|---|---|---|---|
| Intercept | -5.9643 | 2.89999 | -2.0567 | 0.04339 |
| Total_fat_g | 8.98995 | 0.14498 | 62.0078 | 1.4e-63 |
| Carbohydrate_g | 4.0975 | 0.07102 | 57.6913 | 2.2e-61 |
| Protein_g | 4.40133 | 0.39785 | 11.0628 | 4.2e-17 |

Made in the USA
Lexington, KY
06 May 2014